Shadows of a Sunbelt City

Shadows of
a Sunbelt City

THE ENVIRONMENT, RACISM, AND THE
KNOWLEDGE ECONOMY IN AUSTIN

ELIOT TRETTER

THE UNIVERSITY OF GEORGIA PRESS
Athens

Chapter 4 was originally published in different form as "Sustainability and Neoliberal Urban Development: The Environment, Crime and the Remaking of Austin's Downtown," in *Urban Studies* 50, no. 11 (2013): 2222–2227, and online before print March 6, 2013, doi: 10.1177/0042098013478234.

Chapter 5 was originally published in different form as "Contesting Sustainability: 'Smart Growth' and the Redevelopment of Austin's Eastside," *International Journal of Urban and Regional Research*, 37 (January 2013): 297–310. © Urban Research Publications Limited.

© 2016 by the University of Georgia Press
Athens, Georgia 30602
www.ugapress.org
All rights reserved
Set in 10/12.5 Minion Pro by Kaelin Chappell Broaddus

Most University of Georgia Press titles are
available from popular e-book vendors.

Printed digitally

Library of Congress Cataloging-in-Publication Data
Names: Tretter, Eliot.
Title: Shadows of a sunbelt city : the environment, racism, and the knowledge economy in
  Austin / Eliot Tretter.
Description: Athens : The University of Georgia Press, 2016. | Includes bibliographical
  references and index.
Identifiers: LCCN 2015023654 | ISBN 9780820344881 (hardcover : alkaline paper) |
  ISBN 9780820344898 (paperback : alkaline paper) | ISBN 9780820349091 (ebook)
Subjects: LCSH: Urban ecology (Sociology)—Texas—Austin. | Racism—Texas—Austin.
  | Knowledge economy—Texas—Austin. | City planning—Texas—Austin. | Austin (Tex.)—
  Environmental conditions. | Austin (Tex.)—Race relations. | Austin (Tex.)—Economic
  conditions. | University of Texas at Austin. | Universities and colleges—Social aspects—
  Texas—Austin. | Universities and colleges—Environmental aspects—Texas—Austin.
Classification: LCC HT243.U62 A9795 2016 | DDC 307.7609764/31—dc23 LC record available
  at http://lccn.loc.gov/2015023654

CONTENTS

ACKNOWLEDGMENTS

It is hard to thank everyone I have relied on to write this book, but below I would like to acknowledge the generous help and support of a number of people and institutions.

I would like to thank the following librarians for research assistance: Evan Hocker and Roy Hinojosa at the Dolph Briscoe Center for American History; Mike Miller, Gloria Espitia, Molly Hults, Karen Riles, and Daniel Alonso at the Austin History Center; Stanley Fanaras from the National Archives; and Servando Hernandez at the Travis County Records Department. These people provided their time and support and often went out their way to find additional sources of information for me. The descriptive richness of many of the cases studies found in this book can be attributed directly to their assistance.

I also wish to thank the following individuals for allowing me to interview them: Susana Almanza, Bill Bunch, Bill Spelman, Frank Cooksey, Pike Powers, Robert Knight, Will Wynn, Jerry Rusthoven, Mark Yznaga, Glenn West, Mike Clark-Madison, Bruce Todd, Daryl Slusher, Ron Kessler, Oscar Garza, Marcos de Leon, Charles Heimsath, Mary Arnold, Cory Walton, Max Nofziger, and Michael Wilt. I learned a great deal about Austin from my conversations with all of them. Additionally, I thank Michael Kanin, Elizabeth Pagano, Nora Ankrum, and Heidi Gerbracht for connecting me with people to interview and for offering interesting interpretations of Austin events.

Several institutions and colleagues helped in various ways with this project. First, I would like to thank the Department of Geography and the Environment at the University of Texas at Austin, where I was a lecturer for many years. In particular, thanks to Leo Zonn, Kenneth Young, and Sheryl Luzzadder-Beach for supporting my affiliation with that department. In addition to conducting research with that department's support, I enjoyed teaching many classes there, especially The Modern American City. Teaching afforded me the opportunity to put a lot of the research contained in this book into a macro-analytic framework. I would also like to acknowledge the generous grant I received from UT-Austin's

Institute for Urban Policy Research and Analysis, and I thank King Davis and Eric Tang for supporting my research on segregation. Thanks also to the Department of Geography at the University of Calgary, especially John Yackel, and to the Faculty of Arts, which provided some financial support for research. Finally, I appreciate the support of current and former staff of the University of Georgia Press, and including Derek Krissoff, Beth Snead, Mick Gusinde-Duffy, John Joerschke, and Jennifer Comeau. Series editor Nik Heynen has been a real supporter of this project since I first proposed it to him many years ago. Thanks also to Erica Schoenberger and my other, anonymous reviewer, who read an earlier version of this book and provided me with valuable insights and comments.

I would be remiss not to acknowledge the myriad contributions of several professional friends. I am grateful to Richard Heyman, Benjamin Brower, Andrew Busch, Byron Miller, Robert Resch, Roger Baker, Joshua Long, Sarah Dooling, Marie Le Guen, and Bo McCarver for their feedback on earlier drafts of my manuscript. I would also like to give a special thanks to Elizabeth Mueller, who, in addition to always giving me great feedback, provided some of the data used in chapter 5; M. Anwar Sounny-Slitine, who made maps and collected and processed data on the location of high-technology firms and toxic pollution as well as the geography of restrictive covenants; City of Austin demographer Ryan Robinson, who provided shape files and other data used in some maps and charts; Ken Martin and Bob Ward, for allowing me to use maps they produced; Robin Poitras, who created or edited the maps and graphs used in this book; and Daniel Austin Read, for taking the photo for the cover. I am also grateful to the MonkeyWrench Book Collective for allowing me to present earlier drafts of many chapters as public lectures; I benefited greatly from the critical feedback I received afterward. I am also grateful to Roger Keil, then editor of the *International Journal of Urban and Regional Research*, for his feedback on a version of what became chapter 5, as well as Gordon Macleod, Robert Krueger, and David Gibbs for editing a special issue of *Urban Studies* that contains an article largely republished here as chapter 4.

Finally, I would like to recognize the invaluable support I received from lifelong friends and family. I thank my childhood friends Joshua Rosenblatt, for his excellent copyediting, and Francis Cody, William Nelson, and Michael Roller for being intellectual peers since we were wide-eyed teenagers. I would like to thank my parents, Carl and Beryl Tretter, my grandparents, Ruth and George Tretter, and my brother, Robb Tretter, and his family for their support over the last decade. Finally, I want to thank my partner, Nevena Ivanović, who read and edited many chapters and provided endless support and intellectual contributions. I dedicate this book to her.

Shadows of a Sunbelt City

INTRODUCTION

From the dirty brown river at the foot of the Avenue [First Street] to the heights of Mount Bonnell[,] life [in Austin] seems to move with a dream-like quality in which the hum of busy industry is mixed with somnambulant quietness.
Sara Lacy, draft for the Federal Writer's Project Guide to Austin

We live in a prosperous region, but within it are places and people who do not share that prosperity. It is sometimes difficult to call attention to the shadows when the Sun Belt shines so brightly. And given the persistence of the booster ethic and the tendency to emphasize the pleasant and positive, the task can be daunting.
David Goldfield, *Region, Race, and Cities: Interpreting the Urban South*

Openings

In the finale of the hit NBC TV show *The Office* (a remake of the popular British TV series by the same name), viewers learned that three of the show's main characters, Pam Beesly, Jim Halpert, and Darryl Philbin, would be moving from Scranton, Pennsylvania, to Austin, Texas, so that the two male characters could assume their dream jobs at a new small start-up firm. The characters' move is obviously one of upward mobility, given the show's sardonic depiction of their former jobs at the fictional Dunder Mifflin Paper Company and everyday life in Scranton. At one point in the episode, Darryl, talking to another coworker, described Austin as "amazing. It's hot, the music is awesome, and the tacos are for real." Later in another scene, Pam and Jim tried to persuade another character, the socially awkward longtime Scranton native Dwight Schrute, to visit them in Austin. His response was, "For what? The art? The music? The incredible night-life? No, thank you."

What is remarkable about the representations of Scranton and Austin portrayed in *The Office* is how much they reflect and reinforce popularly held opinions about the two cities. While Scranton in the 1920s was a leading heavy industrial center for coal in the United States, today the city struggles to arrest its population decline and, according to one commentator, it "represents the typical declining rustbelt city. Often serving as the butt of jokes" (Rich 2013, 366). In contrast, Austin was barely a city in 1940, but since 1980 its population has grown more than 150 percent. It now represents the typical prospering innovative city. "Everywhere you look," observed Andrew Park, "cities big and small are trying to get in touch with their inner Austin" (Park 2007, 43). Over and over while undertaking and presenting my research, I have found Austin represented as something like the "non-Scranton": Austin is believed to have a great quality of life, to be a dynamic and creative place for new start-up firms and a base for artists and musicians, and to offer sunny weather, great food, and a fantastic nightlife. Certainly Austin is a dynamic city, and I would not argue that Austin is not a great place to live, work, or play (at least for certain kinds of people), but most of this book focuses on Austin's shadows—aspects of the city's history, everyday life, and transformations that have been hidden by the bright light this image casts.

Since the late nineteenth century, Austin has been Texas's state capital and home to the flagship campus of the University of Texas. In the last thirty years it has rapidly become a leading center for innovation in the high-technology, knowledge economy. Its quick climb as an industrial hub is often characterized in Panglossian terms (Florida 2005; Smilor, Gibson, and Kozmetsky 1989; Powers 2004). The story, frequently repeated, is that national and international high-technology firms were drawn to the Austin region because of its reputation for a high quality of life (particularly its low cost of living), its culture (entrepreneurial, bohemian, hedonistic, and/or tolerant), and its aggressive recruitment campaign by an enlightened and farsighted local growth coalition. The result was that practically overnight, a sleepy college and administrative town was rocketed into the center of the cognitive-creative-capitalist economy. According to this depiction, Austin's economic success resulted from a group of *a priori* special qualities and a harmonious and fortuitous relationship among members of a growth coalition (made up of the University of Texas, the state and local governments, and the local business community) that produced few net losses; one business professor from the University of Texas has gone so far as to dub this the Austin Model (Butler 2004, 2010).

I was initially motivated to write this book because I noticed after some research that the characterization mentioned in the previous paragraph about why and how Austin has grown seemed only partially correct. As I stress in varying ways throughout this book, the following two correctives need to be added to this account: (1) the region's industrial and urban development has come at a high cost, at least to some communities, and (2) there has been significant conflict among some members of the local growth coalition. Moreover, it was not so much preexisting local factors as the substantial reworking of local conditions, particularly political and spatial arrangements, that helped propel the city's growth. I investigate how the University of Texas attained a much stronger role in the local growth coalition and how leading factions from the environmental and business communities became entwined in a common growth agenda. Furthermore, I point to the significance of UT's land development programs, both around the main campus and in other parts of the city, and how past planning decisions are important in attempts to promote more growth in Austin's central areas.

Over the course of my research, I began to notice that the dominant academic accounts given for the relationship between the knowledge economy and regional development patterns in the United States missed one central factor. Nearly everybody, including me, suggests that some small- or medium-sized cities, like Austin, have been able to emerge as significant regional centers of growth within the global economy because of the presence of a university (or a group of universities) and their vital role as the backbone of regional innovation networks. Several common reasons given in the literature for why universities have this remarkable role are that they produce lots of skilled labor, help encourage the development of spin-off firms, raise an area's reputation, and create a desirable cultural sensibility. Remarkably, none of these accounts addresses the issue of universities as land developers, which is the most essential factor I gleaned by examining the historical record (and this was before I had read Margaret O'Mara's remarkable book on the topic). In fact, what stands out in Austin is how large a role the federal government (in conjunction with the municipality), and then later the State of Texas, had in creating the industrial landscape in Austin, primarily because they empowered the university not only as an institution of higher learning, but as a land developer. This fundamental role of universities changes how we understand the knowledge economy's development because it stresses how significant the practices remaking the urban environment are for producing certain economic activities.

While reading the literature on cities and the knowledge economy, I was also struck that rent-seeking practices, as they did in the past, remain a prime mover of urbanization. In particular, I noticed how the fortunes of contemporary urban growth coalitions, like so many previous kinds of urban-based political alliances, depend on economic activities that secure a steady stream of revenues from (legally warranted) monopolies (Braudel 1982). Certainly others have observed how the "monopoly powers of place (expressed in place-specific process and product configurations)" are untradeable assets that significantly enhance a city's competitive advantage (Scott 1997, 325). More recently, there has been recognition that the growing uneven development among cities is strongly connected to the revenues that can be received by maintaining the high exchange value of technological and what I will call "knowledge" rents (Storper 2013). Yet as far as I know, there has not been a concerted effort to connect the significance of revenues from rent seeking as a development driver to the fundamental role universities have attained in some urban economies. This emphasis changes our understanding of urban development because it links the growing significance of universities in national economies to the development of new forms of monopoly power.

Like so many people interested in how and why cities change, I was also driven by a perennial question in urban studies: who governs a city? While urban governance is an enormously broad topic, I was primarily concerned with urban planning and industrial policy. After several years of research, I began to see the relationship between the present efforts to rework Austin's spatial form and the specific qualities of the city's contemporary governing coalition. In particular, I began to think about how the City Council's recent policies and programs to refashion certain areas of the city, under the rubric of Smart Growth, were affecting communities whose priorities were not central to any dominant faction of the governing coalition. Later, examining the relationship between political coalitions and spatial arrangements over a much longer period of time, I discovered interesting parallels between the past and the present.

More precisely, however, I wanted to connect my research program about Austin to debates about urban governance that were unfolding in the emerging field of critical urban sustainability studies. Unlike other approaches to sustainability, this body of scholarship uses central concepts from urban political economy to understand the evolution and impacts of policies and programs designed to promote urban sustainability (Keil 2003; Krueger and Gibbs 2007; Heynen 2013). In terms of governance, this research has focused on how selectively incorporating urban sustainability principles, especially in terms of the natural environment, has become an essential part of contemporary urban entrepreneurialism and of the ways cities gain a competitive advantage in the interurban struggle for investment. Contemporary planning practices figure strongly in these efforts, and while they are able to meet some goals for better environmental stewardship, in most cases these urban planning efforts are unable to similarly improve social equity and may even make cities more unequal. Certainly one unique aspect of this literature is its emphasis on the significance of spatial relations, especially how the remaking of urban spaces is essential to fulfilling urban sustainability goals. However, the reorganization of a city's spatial form, especially in defined areas, can have a bigger role in urban governance than just being important for a city's competitive strategy. In Austin, the adoption of particular planning and design principles, and their implementation to refashion specific areas of a city, helped resolve tensions among the political factions competing for local government control. This outcome tells us that attention to spatial factors must remain at the heart of critical sustainability studies.

Shadows of a Sunbelt City is strongly rooted in a radical tradition of urban geography; therefore, it is worth mentioning that the three themes—the environment, racism, and the knowledge economy—used throughout the book are not simple substitutes for the more standard categories of ecology, society, and economy. Instead, the term *environment*, in this book, refers to the entire urban environment, which contains both humans and nonhuman species. Furthermore, *racism* indicates how uneven relationships among different ethnic groups, especially those groups marked as nonwhite or partially white, have affected how the city developed. In particular, I stress how historical legacies and contemporary practices of prejudice and violence have influenced the vulnerability of different social groups to changes to the urban environment. Lastly, *knowledge economy* refers to more than private industrial service firms with high concentrations of human capital; it also includes the university's central function, particularly in its role as a land developer, in the region's economy. Baldly, this book's main argument is that the developing knowledge economy changed and interacted in complex ways with Austin's environment and its system of racial relations; the impact of these transformations has been strongly influenced by a historically varying, but still relatively stable, system of asymmetrical power relations that has engendered both the uneven development among neighborhoods and relative inequalities among peoples.

The evidence I use to make my arguments comes from several sources. Some primary material originates from numerous semistructured interviews I conducted with former and current mayors, City Council members, members of the Greater Austin Chamber of Commerce, real estate developers, administrators in several divisions of the local government, members of various environmental organizations, journalists, members of neighborhood associations, and political activists. Other primary sources such as government reports and institutional records can be found in the following archives and libraries: the Austin History Center; the Dolph Briscoe Center for American History; the Perry-Castañeda Library; the Tarleton Law Library; the Texas State Library and Archives; and the National Archives. Finally, I relied on newspaper and magazine articles, especially from the *Austin Business Journal*, the *Austin American-Statesman*, the *Austin Chronicle*, the *Daily Texan*, and the *Alcalde*.

Outline of the Book

This book is divided into two parts and six chapters. It is not written to reflect a straightforward historical narrative, although chapters 2 and 3 could be read in this way and chapters 4, 5, and 6 are all about contemporary development in Austin.

Part 1 primarily deals with the knowledge economy and more specifically the changing role of universities as drivers of urban and regional development.

Chapter 1 situates the development of the knowledge economy in Austin within the broader literature in urban political economy. On the one hand, the chapter explores how aspects of Austin's development can be accounted for by two dominant schools of thought about what propels regional growth and un-even development. On the other hand, I argue that these schools of thought are inadequate, and I suggest an alternative account grounded in an understanding of spatiotemporal dynamics of the circulation of capital in and out of three different circuits. I draw special attention to what geographers call the secondary circuit of capital (i.e., investments into physical infrastructures) and also the tertiary circuit of capital (i.e., investments into social infrastructures). I use these concepts to understand why science and technology have a more central role in the economy (especially their part in research and development) and how this relates to changes in the economic value of university research and the growing financialization of R & D. I also note how two forms of rent, technological and knowledge rents, are increasingly important to urban growth, and I draw specific attention to the importance of these forms of rent seeking for universities in general and for the University of Texas at Austin in particular.

Chapter 2 is the first of two chapters to look at the role of the University of Texas as one of Austin's chief land developers. This chapter's focus is on an urban renewal initiative that was proposed and completed in central Austin in the late 1960s and early 1970s. The chapter highlights how the reasons behind this substantial land-acquisition program were different from those of previous territorial expansions, and how the 1960s effort was a central part of the university's transformation into a nationally ranked research university. I also describe how the City of Austin and its agent (the Austin Urban Renewal Authority), the Board of Regents of the University of Texas, and the federal government acted in concert because of a shared interest in redevelopment and promoting the university's expanding role in the local economy. However, there were social costs associated with this expansion that were, by and large, borne only by more modestly resourced African Americans in neighborhoods adjacent to the university.

The focus on the central role that the University of Texas played as a land developer in Austin continues into chapter 3, but the investigation moves to the 1980s. Unlike chapter 2, which emphasized the central role of federal policy, this chapter homes in on the relationship between the university and the government of the State of Texas. I focus on how the state government helped further transform the university's role, as its agent, into a land developer that could support the industrialization of central Texas. The university's actions, I believe, are unusual for the way the university used its unique bonding authority (granted under the provisions related to the Permanent University Fund) solely for the purpose of obtaining and developing an industrial site for two national corporate high-technology consortiums, the Microelectronics and Computer Corporation (MCC) and Sematech. Capturing this investment helped reposition Austin within the national, and later global, urban hierarchy and, I argue, led to the birth of a new configuration of actors in the local growth coalition.

Part 2 shifts to a discussion about contemporary planning practice in Austin, particularly Smart Growth, and its relationship to the city's urban governance.

Chapter 4 takes a critical look at Austin's turn toward urban sustainability and particularly how the sustainability focus relates to the city's strategy of revitalizing its existing urban core. This chapter engages directly with growth-machine theory but argues that the growth-machine thesis must be amended to include the concept of the "sustainability fix" and its emphasis on urban competitiveness and entrepreneurialism, as well as a concern with policing and programs of social control. I argue that the political compromise reached between the growth and antigrowth coalitions became workable because the latter were willing to accept redistributing the costs of growth away from "nonhuman species" and onto humans, particularly homeless people.

Chapter 5 builds on this thesis in the context of how the remaking of Austin's urban core is affecting the city's more modestly resourced communities of color. I concentrate on the tensions that existed in Austin between the "environmental" and "environmental justice" communities. I pose the question of how a selective environmental sustainability framework was effectively incorporated into the hegemonic vision of Austin's strategic growth plan. The answer, I believe, can be found in asking what counts as "the environment" for environmentalists. I show that the present political arrangements between environmentalists and the business community were born out of a compromise stipulating that environmentalists would support remaking Austin's downtown if the business community supported preserving an area in the suburbs for nonhuman species. The unstated assumption of both sides was that the transformation of the downtown physical and social environment (especially the displacement of communities of nonwhite minorities) was not an environmental problem because environmental concerns were limited to internalizing the effects of urbanization on nonhuman species.

Finally, in chapter 6, I continue to investigate urban planning and governance but looking at a much longer period of time. In particular, the chapter connects comprehensive planning efforts in Austin and a number of reforms of the formal systems of electing local government officials. As in previous chapters, I argue that studying this relationship reveals how much the business community's priorities influence the city's urban governance. I draw out parallels between the past and the present. One hundred years ago, official reforms of municipal representation clearly influenced the priorities established in the city's first master plan. I argue that this correspondence is maintained in future attempts to update Austin's comprehensive plan and enact further reforms in the city's system of political representation. In this chapter, I emphasize what can be gleaned by looking at the period from the 1970s to the first decade of the twenty-first century, when both efforts faltered.

PART I

Forces of Growth

CHAPTER 1

The Making of a Globalized Austin

The Environment, Racism, and the Knowledge Economy

Down the standard path, capital is put into a production process, converted into a commodity and sold upon the market under the tight discipline of socially necessary turnover time. But money can also flow into fixed capital and consumption fund formation, including the formation of physical infrastructures. It can also flow into science and technology, [into] improved administration, or into the creation and maintenances of a variety of social infrastructures which enhance the conditions for surplus value production. The temporal discipline down each of these paths is much relaxed because the turnover times are much longer.

> David Harvey, *The Limits to Capital*

Because of its central location in the state and its convenience to the Industrial Southwest Area, its position as an educational and governmental center and its consequent supply of highly trained professional people and technical specialists required by certain high performance industries, and extraordinar[il]y desirable living conditions which have attracted and held an efficient labor force, Austin probably should be attractive to new industries such as research, fabricating plants, electronics manufacturing, precision tool and instrument manufacture and other "light" industrial operations.

> City of Austin, *Austin Development Plan*

In 2008, when the Globalization and World Cities (GAWC) Research Network ranked cities in the global urban hierarchy, Austin placed in the category of "sufficiency" or, more precisely, in the bottom of half of the fifth-tiered cities. Two years later, in 2010, Austin had climbed two tiers in the GAWC rankings to the lower third of the "gamma" group, but in 2012 it was demoted to the upper echelons of cities characterized as having only "high sufficiency." The GAWC metric is "based upon the office networks" of leading "advanced producer service firms," and although this metric is not without problems, it is still telling that Austin, which had not even made the first ranking in 2000, was by 2008 fully recognized to be ensconced as a medium-sized city-region within the global urban network (Network 2012).

This chapter's central purpose is to place Austin's patterns of urban and regional development since the 1970s within a much larger political, social, and economic context and to show some of the factors that have propelled the city into the middle ranks of global urban hierarchy. While others have offered accounts of the city's rapid ascent, what I suggest is that Austin's rise into the global marketplace results from how its growth coalition, because of a unique set of local conditions, was able to effectively capitalize on the growing flow of investment into what geographers call the "tertiary circuit of capital." I argue that this rising stream of investment, which helped propel Austin's regional growth, was driven in part by the internal spatiotemporal dynamics of the tertiary circuit (its long turnover time and ability to absorb tremendous amounts of investment) and by external developments—that is, the development of legal and financial systems designed to support a system of knowledge-rent seeking.

This chapter has four sections. The first two outline three longstanding and important issues that affect urban governance in Austin and provide a brief history of the city's urban development and integration into the global economy (points developed in more detail in other chapters). In the third section, the focus shifts to the knowledge economy and the two most dominant schools of thought about what propels regional growth and uneven development, and I suggest an alternative account grounded in an understanding of spatiotemporal dynamics of the circulation of capital. The fourth section outlines many important structural changes in the U.S. economy over the last forty years and highlights their impact on regional development and the urbanization process. I draw special attention to the increasing financialization and internationalization of the U.S. economy (particularly for real estate) and discuss what geographers call the *secondary circuit of capital* (investments into physical infrastructures). Finally, I turn to a discussion of science and technology and their role in research and development. Here, I discuss the changing role of university research and the growing financialization of R & D, grounding it in a discussion of the *tertiary circuit of capital* (investments into social infrastructures).

Twentieth-Century Austin:
Urban Governance, Power, Racism, and Inequality

As in other Texas cities, during the twentieth century a progrowth coalition of members from the Austin business community, represented by its principal organization, the Chamber of Commerce, substantially influenced how the city was governed and, in turn, its urban planning priorities. While their dominancy has not gone uncontested, as other chapters in this book demonstrate, the

Chamber's ideals about what kind of city Austin ought to be are still reflected in the vision that is largely embraced by the city's governing elites.

Like other Texas cities, Austin has a history going back to the mid-nineteenth century that is marked by social, cultural, and legal systems that supported white supremacy and "nonwhite" servitude (Lack 1981). As Austin continued to grow, its employment structures, system of education, relations of social intercourse, housing patterns, and other factors were organized by both legal and extralegal means to ensure racial inequalities and white domination (McDonald 2012). In the twentieth century, this racial hierarchy built on a strict white/black binary was complicated by the urbanization of a large and growing Hispanic population, who were white by law but in practice were treated as something other than "white." While the structures of systemic racism have changed during Austin's history, as other chapters in this book show, the legacies and realities of this unequal system continue to influence the city's governance, patterns of growth, and urban planning priorities.

During the twentieth century, wealth inequality has been considerable in Austin. While the disparities in income and wealth between white and nonwhite have been substantial, even among "whites" there have been significant class differences. For instance, in 1938 Lyndon Baines Johnson, then representing Austin in Congress, complained in his famous "Tarnish on the Violet Crown" speech that the city had a higher percentage of homes that were in disrepair, uninhabitable, overcrowded, or without adequate facilities than sixty-four other cities across the United States (Johnson 1938). Moreover, a survey conducted in the 1950s by famed urban economist Wilbur Thompson ranked Austin the fourth most unequal city in the United States, a situation that Thompson suggested was characteristic of cities without large manufacturing bases and with histories of discrimination toward African Americans (Thompson 1965, 110). Certainly, economic inequalities have risen in Austin since the 1980s (as they have in nearly all metropolitan statistical areas [MSAs] across the country), and class differences among social groups have become more apparent and accentuated by spatial distribution and segregation inequality (Straubhaar et al. 2012). Nevertheless, Austin was already a city with significant differences in wealth before the contemporary development of the knowledge economy, and the increasing stratification of the technical division of labor has helped reinforce and deepen existing patterns of wealth inequalities.

New Developments, Old Patterns:
A Brief History of Austin's Urbanization and Its High-Technology Industrialization

Beginning in the late 1970s and accelerating rapidly in the late 1980s, Austin emerged as one of the most competitive midsize cities in the core of the global economy and one of the fastest-growing urban areas in the United States (Long 2010; McCann 2003; McCann 2007). In general, Austin's growth reflects the trend of other Sunbelt states that have had substantially faster rates of urbanization over the last forty years than other areas in the United States (Storper and Walker 1984; Shermer 2011). Regionally, in Texas's urban triangle—formed by the mega-conurbations around Dallas–Fort Worth, San Antonio–Austin, and Houston-Galveston—there has been sustained, rapid, and intensive interurban growth over the last four decades (Davies 1986). Although primary-sector activities, such as raw mineral extraction and agriculture, still have a considerable role in Texas's economy and are vital in sustaining its high rates of urbanization, there has been a significant diversification of the state's economy since the 1980s, and high-technology industries in tertiary or quaternary sectors such as software design, semiconductor manufacturing, aerospace, biotechnology, and computer equipment have become important engines of economic growth in urbanized areas (Rodriquez and Fukasawa 1996; Lyons and Luker 1998).

But Austin's story of economic and urban development is nevertheless unique. While the Houston and Dallas regions have ranked among the largest cities in the United States since at least the 1960s, the Austin metropolitan area has been one the ten fastest-growing regions in the country over the last thirty years, far outpacing the rates of urban growth in other Texas cities (Frey 2012). Moreover, unlike these regional counterparts, Austin emerged as a center in the global economy without a significant manufacturing or industrial base (Feagin 1988; Kaplan 1983; Melosi 1983; Phillips 2006; Shelton et al. 1989).

Since at least the beginning of the twentieth century, Austin's urban development has benefited from the city's roles as the state capital and home to the flagship campus of the University of Texas. By all accounts, Austin's governing elites attempted to exploit being the exclusive location of these institutions in order to promote growth, and for decades the city remained overwhelmingly dominated by these institutions and their outsized role in the local labor market. While Austin's local growth coalition embraced the city's role as an educational and government center before the 1950s, they still tried, though mostly unsuccessfully, to promote Austin's manufacturing and commercial base (Austin Board of Trade 1894; Austin Chamber of Commerce 1916). Despite the fact that the state government and the university remained the essential ingredients that propelled Austin's growth in the latter half of the twentieth century, industrializing central Texas became possible only after several substantial changes in the structures of the economy.

Since at least the 1960s, in part because its two main employers, the state government and the university, dominated the local job market, Austin has been distinguished by the proportion of its population with a high level of formal education. A 1966 study on Austin's social and neighborhood characteristics noted that in 1964 the city's "median education level . . . was estimated to be 11.8

years"—a figure the authors claimed was one of the highest education indexes for any "medium-sized" U.S. city (Hazard, Kelsey, and Strickland 1966, 195). In 2000, Austin ranked second among seventy-five cities in the "proportion of adults holding at least a bachelor's degree" (Gottlieb and Fogarty 2003, 329). Recently, Jeremiah Spence and others have argued, the high level of educational attainment in Austin has been a significant driver of the city's competitiveness, but the disparities in formal education are also increasingly important in creating a more socially bifurcated city (Spence et al. 2012).

Furthermore, Austin has had a distinctive reputation for the countercultural movement that developed there after the 1950s and for its "liberal" political and cultural atmosphere, and the city has continued to be seen as distinct from other Texas cities (Long 2009; Rossinow 1998). Again, the university's elephantine role is the most significant contribution to these twin developments. Richard Florida has contended that over the last thirty years, Austin's tolerant and alternative sensibility has been a key driver of growth in Austin, as it has created a unique mise-en-scène and enhanced the city's quality of life, making Austin more attractive to young, upwardly mobile professionals and high-technology firms (Florida 2005). Although such an account does point out the importance of cultural traditions and other amenities or extraeconomic factors that have supported the city's growth, it largely discounts the significance of other political-economic factors that may have been more important in driving the region's competitiveness.

Austin has unique patterns of residential segregation, which have remained relatively fixed since they were established in the first half of the twentieth century, despite the city's population growth. Wealthier residents reside mainly in the west and northwest, while the city's more modestly resourced communities live in the south and southeast. As more technically skilled people have come to the city, particularly those seeking high-paying jobs in the growing high-technology sectors, they have settled in existing neighborhoods or in new developments in the north or northwest (Orum 1987, 306–348). In contrast, the neighborhoods in the eastern and southeastern areas—due to a number of legal, extralegal, and financial constraints that are explored in more detail in chapter 6—have developed as the principal residential areas for more modestly resourced communities. Although recently, as chapters 4 and 5 discuss in detail, wealth and racial demographics in the eastern areas have shifted because of gentrification (especially in those historically more modestly resourced neighborhoods located near the city center), even today longtime and newly arriving nonwhite minorities and less technically skilled laborers tend to live in East Austin (Skop and Buentello 2008; Wilson, Rhodes, and Glickman 2007).

In the 1980s, Austin's urban geography was significantly affected by a land-development boom and a later bust, both driven in part by a lax lending market and in part by the development of an infrastructure of high-technology firms. By the early 1980s, Austin had attracted the investment of a number of high-technology firms, which had spawned some development in the city's suburban fringe. However, when the Microelectronics and Computer Corporation (MCC, a firm discussed in detail in chapter 3) announced its plan to build its new headquarters and facilities in a northwest suburb, a land boom began, anticipating the arrival of legions of new residents. In a frenzy, developers drew up plans for large-scale residential and commercial projects in the areas adjacent, or easily accessible, to MCC's site (Kahn and Farley 1984, 82). These projects found easy financing in the loose lending market that had been created in Texas when the high price of oil in the early 1980s produced large reserves of money in the state's economy (Feagin 1988). Banks, and particularly newly deregulated savings and loans swimming in cash, were eager to finance commercial and residential projects in Austin's newly developing suburbs and its older downtown (Barna 1992). By the mid-1980s, Austin's "financial institutions invested 50 to 75 percent of their portfolios in land and construction" (Kim 1998, 79). However, in 1985 the price of oil fell and the pace of urban growth slackened. Developers found that they were unable to lease space in their new office buildings or sell their land holdings. As a result, some developers and bankers, among them some preeminent local and state figures, took huge financial losses; others, unable to meet their debt obligations, went bankrupt. For instance, former Texas governor and secretary of the treasury John Connally lost millions on a new high-end residential development in a western suburb of Austin, while local businesspeople John Watson and Pike Powers, who had also speculated in the northwest, were forced into bankruptcy (Hight 1994; Gibson and Rogers 1994, 438–439; Copelin 1996).

In the 1990s, a boom dynamic returned, this time fueled by the so-called tech or dotcom bubble, which significantly propelled Austin's industrial development and its fast ascent into a leading position in the global economy. As Long points out, "Between 1989 and 1999, over three hundred companies—most of them tech-related—had either headquartered or located offices in Austin" (Long 2010, 39). Moreover, in the 1990s Austin was, for the first time, able to attract larger amounts of investments from venture capitalists (Oden, Kang, and Kwon 2007). Although there were some venture capital firms based in Austin in the 1980s, these firms did not invest in Austin, nor was the city able to draw much outside venture capital (Smilor et al. 2007). By the mid-1990s, however, Austin began to receive about "one to two percent of the [United States'] total . . . venture capital investment" (about $2.3 billion), and "the rate of growth" of this investment in the city was "almost double the rate of growth" for the United States (Mahdjoubi 2004, 149). For example, Austin's largest and most significant venture capital firm, Austin Ventures (founded in 1984), financed 26 new Austin businesses in 1996 and 166 by 1999 (Walker 2000). In the late 1990s, Austin was again swimming in money; at the height of the technology bubble, according to Walker, eleven companies formed in Austin in 1999 had an estimated stock market value of $23 billion. But after the bubble burst in late 2000, many of the businesses Austin Ventures had helped finance (and many other firms) went bankrupt, the total stock market value of those eleven firms dropped by about 45 percent, and Austin lost an estimated forty-one thousand jobs (Peng 2007). However, the bust in 2000 had a more muted impact on Austin's real estate market than the collapse more than a decade earlier, and overall the crash had less effect on slowing Austin's urban growth; during the first decade of the twenty-first century, Austin continued to expand its profile as a center of high technology, attracting substantial investment from several more multinational companies and venture capital funds (Florida 2013; Gibson and Butler 2013).

The increasingly significant role played by high-technology firms in Austin since the 1970s has been accompanied by the increasingly significant role of the University of Texas (with the backing of the state of Texas) in supporting the region's industrialization. The university's support for undergraduate and graduate education (especially in the material, natural, and physical sciences) yielded a substantial pool of technically skilled workers who could support the endogenous development of high-technology firms. Moreover, UT faculty and researchers have founded several successful technology companies, and the university's land-development programs, as chapters 2 and 3 show, have had an essential role in supporting the development of high-technology firms in the city. More recently, the university's expanded role in supporting knowledge transfer to industry and the commercialization of its own discoveries has proved critical for Austin's regional development. For instance, between 2003 and 2011 UT issued "276 U.S. and 148 foreign patents" and its annual revenues from licensing income increased from about "$500,000 in 1992 to over $25 million in 2011," and in the years since 2003, "58 spinoffs [have been created] based on UT Austin research" (Gibson and Butler 2013, 69).

Producing a Knowledge Economy:
The Tertiary Circuit, Urbanization, and Knowledge-Rents

Since the 1970s, the monetary returns from intellectual property holdings have become more significant for sustaining economic growth, especially in highly industrialized regions of the world (Greenhalgh and Rogers 2010). As a consequence, the legal infrastructure, designed to ensure that these revenues can be collected, has been deepening and broadening because knowledge that is alienated into products and processes is a "nonrivalrous good" (a good whose use by one person does not preclude the use by another) and can be easily stolen or copied (Perelman 2003). On the one hand, new legal mechanisms have been created to expand and enhance enforcement over intellectual monopolies and limit the unrestricted proliferation of new discoveries. On the other hand, new frontiers for economic opportunity have been opened because of the expansion of intellectual property into new domains such as the life sciences, and new legislation, treaties, and court rulings have influenced how these domains can be commercialized (Berman 2011; Zeller 2007). Without the development of a more robust legal infrastructure to preserve the vitality of these intellectual property holdings, investments into many types of R & D, especially those that involve the commercialization of basic research, would not be financially attractive because there would be no way to ensure a reasonably good chance of capturing a stream of viable technological or knowledge-rents (Teixeira and Rotta 2012).

Patterns of urbanization and industrialization in the second half of the twentieth century reflect the growing significance of technological and knowledge-rent seeking and, I suggest, the spatiotemporal qualities of investment flows into what geographers have called the *tertiary circuit of capital*. Certainly, many scholars have shown how the larger shift in the U.S. economy in the 1970s resulted from the inability of the economic system that developed after the 1940s—a system that relied on increasing returns to economies of scale—to effectively resolve a myriad of spatiotemporal tensions that it created (Schoenberger 1997). In particular, scholarship (described in more detail in the next section) has looked at how the development of new technologies, which restructured core economic sectors by making the production process faster, more flexible, and more responsive to consumer and producer needs and desires, helped resolve these tensions. Yet the significance of investment in science and technology, particularly in knowledge production, has remained largely outside this discussion, despite the fact that since the 1970s a growing share of the economy has developed because of a legal system that supports taking rents from intellectual property holdings. Instead, two different, although not unrelated, approaches dominate how scholars explain the forms of urbanization around knowledge production.

One approach is seen in the work of economic geographers who, in exploring the development of new competitive regions, have focused on the significance of the economic returns to social and physical infrastructures and their role in helping firms secure an above-average rate of profit in their sectors (Moulaert and Sekia 2003). Perhaps the most clearly stated version of this position appears in the work of Michael Storper and Allen Scott, who have argued that extra-economic factors (untraded interdependencies, cultural milieus, knowledge systems, returns to physical infrastructures, and so on), which are territorialized over many years, become unfungible assets that grant certain places a competitive advantage (Scott and Storper 1992). On the one hand, these infrastructures allow firms to exploit the external economies of scale and lower transaction costs. On the other hand, these same infrastructures create high barriers for new regions attempting to enter the market. Moreover, the returns from the ability to protect and maintain the "technological rents" accrued by certain firms or industrial districts significantly drive regional competitiveness, innovation, development, and performance (Storper 2009).

Another approach, recently advanced by Margaret O'Mara, puts a stronger emphasis on political structures, particularly the federal government's policies and the strategies of local growth coalitions, to account for the development of successful regions, especially those with a strong research university (O'Mara 2005). Certainly, the federal government's uneven geographical spending patterns have had a strong role in the development of high-technology sectors. In particular, the regions that have received (or were able to capture) large federal government expenditures, especially for military-related research, have seen significant positive spillovers that have helped spawn the development of high-technology firms (Leslie 1993; Markusen 1991). Moreover, federal policy has encouraged more university involvement in urban development and in collaborations with private-sector firms. Astute growth coalitions, able to exploit this growing federal and university involvement in urban policy, have sometimes been able to make or remake their cities into profitable centers for knowledge production. More recently, the federal government has helped create a stronger legal framework to make it easier for universities to benefit financially from the commercialization of their discoveries, especially in emerging fields of research in the natural sciences (Berman 2011). As in the past, local growth coalitions have tried to connect the expanded commercial role of university research programs to efforts that promote regional development.

Both of these accounts reveal significant drivers of urban development, yet in both, the spatiotemporal dynamics of capital circulation among three different but related circuits, and the impact of those dynamics on the urbanization process, remain undeveloped. What I argue is that the marked shift of financial firms (especially venture capitalists) toward investment in science and technology, especially into basic research and its potential for commercialization, reflects a propensity to invest in activities whose hallmarks are a slow turnover time before the initial investment can be repaid in the form of captured rental payments. In this regard, science and technology share some basic features with real estate, at least in terms of the spatiotemporal qualities of investment and how earnings are returned in the form of rents. As I will show, like real estate, science and technology that has no proven record of successful commercialization has become a much more substantial area for private financial speculation since the 1970s, even if the research is conducted inside university labs, many of which remain publically supported. What I contend is that cities of knowledge such as Austin, and their growth coalitions, strongly supported by federal policy, have been successful because they were able to capitalize on switching capital into the tertiary circuit and an expanded infrastructure that supports knowledge-rent taking.

Switching Circuits:
The Economic Crisis of the 1970s, the New Economy,
and the Secondary Circuit of Capital

Over the last thirty years, a vast and ever growing body of literature has debated the causes and consequences of the dramatic shifts in the economic structure of the United States since the 1970s. While the contours of these debates are wide, there are some broad areas of agreement among many scholars (Dicken 2007; Webber and Rigby 1996; Brenner 2003; Held 1999; Castells 2011; Sassen 2011). First, financial services and ancillary producer services such as real estate and insurance have significantly expanded. Second, the shift came about because of a crisis in profitability, not productivity, among major industrial firms. Third, technological innovations increased the speed and flexibility of the systems of production, consumption, organization, and distribution, and these changes helped destabilize—in some cases destroy—the competitive advantage shared by some firms that relied on increasing economies of scale to continually lower their costs. Fourth, information- and knowledge-intensive activities became the leading sectors of the economy, and this shift had a tremendous impact on the structure and composition of the technical and social division of labor. Fifth, the labor market underwent a significant restructuring. It became much more "feminized"—that is, many more women entered the workforce—and much more flexible and less secure. Whole segments of the workforce that had been relatively stable, particularly professions dominated by white working-class men, were decimated, and the new jobs that were created were unable to absorb these losses. Sixth, the national state economic policy, broadly known as Keynesianism, failed to adjust to these changes, contributing to what is known as the stagflation crisis in the 1970s, and was largely abandoned in favor of a state fiscal policy known as monetarism and a political program of neoliberalism in the 1980s. Finally, there was an increasing "internationalization" of the U.S. economy, a process facilitated by the substantial deregulation of global trade. Some U.S. firms became more vulnerable to foreign competitors while others grew and became multinational firms. Moreover, new regions in the United States became open to foreign investment, while U.S. firms invested more readily abroad. However, the most visible change associated with the new globalized economy was the loosening of regulations on the flows of finance and the enlargement of the global financial system to facilitate the smoother and more rapid transfer of investment across international borders.

Patterns of urbanization in the United States, like in other periods of history, were affected by these structural changes in the economy. First, financial and business-service firms became more important both as employers and as commanders of greater portions of capital, and the restructuring of older central business districts to support these services reflected this growing economic power and significance. Second, major industrial and commercial firms that were unable to restore their profitability collapsed, and those that were able to did so at the expense of their workforces and physical plants; as a consequence, some industrial cities and regions suffered from excessively high rates of unemployment and wave after wave of plant closures. Moreover, the places that grew around the newly burgeoning sectors of the economy were often far from the deindustrializing areas, and where they were not, costly efforts—often at public expense—were needed to remake the existing landscape to meet the demands of the growing sectors of the economy. Furthermore, wealth inequalities increased in the United States as a whole and especially in metropolitan regions, supported by the technical stratification of employment in the fastest-growing sectors of the economy; their workforces tended to be either highly paid and technically skilled or poorly paid and unskilled. Finally, like national policy, city and state policy increasingly relied on abandoning state welfare policies and on entrepreneurial and market-oriented solutions that sought to encourage job growth by partnering with private enterprise or pursuing strategies to entice inward private investment, especially from large multinational and foreign firms.

Furthermore, U.S. real estate was appreciably financialized and internationalized. Regulatory reforms allowed for changes and innovations in the types and techniques of financial securitization (e.g., financial products, such as promissory notes on real estate mortgages, that could be sold as assets on secondary markets to investors), and these new types of financial derivatives allowed mortgage holders, such as commercial banks, to accelerate the turnover time on their debt holdings. The new markets that developed for trading these debt obligations grew rapidly as domestic and foreign investors found that hefty returns could be earned from these financial vehicles. As a consequence, financial institutions became increasingly willing to lend or leverage more and more of their resources for land development. Moreover, new types of financial firms emerged—for example, Real Estate Investments Trusts (REITs), which pooled money raised from public or private shareholders; bought real estate holdings, such as office buildings; and paid investors dividends from the earnings on their real estate investments (Gotham 2006). Other financial firms, such as insurance companies, pension funds, and investment banks, also increasingly held real estate assets directly, indirectly through REITs, or through their derivatives (securities) in their financial portfolios.

David Harvey's pioneering work on the urban process under capitalism has connected these changes in real estate and finance to the politics and patterns of urbanization. For accumulated capital to be productive, to find profitable outlets, it must be reinvested, but in some cases reinvestment opportunities are limited by the productive capacity in some sectors of the economy. Nevertheless, investing in the built environment may still appear to be an attractive opportunity for at least two reasons: (1) the absorption of large amounts of capital; and

(2) the unique temporal dynamics of such investments (there is slower turn-over between the initial investment and the expected future returns). For these two reasons, the modern financial system has a unique place in the heart of the modern system of landownership. On the one hand, financing allows the future to be pulled into the present—that is, vast sums of capital can be immobilized in land today only because they are borrowed against the promise of a better tomorrow. On the other hand, financing allows the present to be pushed into the future—that is, the promise of earning better returns tomorrow can happen only because vast sums of capital can be immobilized in land today.

It is important to stress that in Harvey's view, in a system of expanding re-production, capital circulates in and out of different circuits. The primary cir-cuit consists of the production and consumption of the "technical and social organization of the work process and the quantities and capacities of the labor force" (Harvey 1999, 408). Profits created out of the *primary circuit* (the primary production process) can be productively reinvested back into the primary cir-cuit—that is, put back into the immediate production process by buying more inputs or increasing a firm's labor-power capacity. But capital can also be pro-ductively moved into at least two other circuits that help facilitate the expanded reproduction of profit.

The *secondary circuit*, to Harvey, consists of productive investments into a fixed capital fund (composed of producer durables and the built environment) or nonproductive investments into a consumption fund of physical infrastruc-ture. The productive flows of capital in and out of the primary and secondary circuits are related to what Harvey calls the differing "amortization times" of investments; investments in fixed capital have high sunken costs, and there is a long turnover time before the initial investment can be recovered from the advances provided to the production process. Moreover, the productive invest-ments firms make into the secondary circuit—that is, into technical and phys-ical improvements in machinery, buildings, or other physical infrastructure—may improve a firm's productivity and allow it to gain a small financial benefit or a rent that comes from an exclusive advantage. These rents may be essential to a firm's competiveness because they increase the firm's rate of profit above the social average for their sector. Although making investments into the secondary circuit may be central for firms to gain or maintain a competitive advantage over their rivals, firms often lack the financial capacity to make new productive investments. Credit and financial markets help fill this gap by providing funds that allow firms to make the needed investments today by leveraging the prom-ise of better returns tomorrow.

But finance also has a special place in real estate because the earnings that come from investments into the improvement of land share some significant qualities with the revenues derived, as interest, from the advancing of money. In particular, both kinds of activities are what Harvey calls "fictitious" because the revenues earned in both practices come only indirectly, from *expectations* about future gains. In the case of land, the asset's value is based on anticipated returns either from benefits on increasing returns or from interest payments (rents or appreciation) that are expected to be earned at some time from the lease or the sale of the land. For Harvey, therefore, the revenue from landowner-ship, whether urbanized or not, is already a form of finance, as it functions as an advanced payment on other endeavors.

In Harvey's view, then, the financialization of real estate was a response to the economic crisis of the 1970s, as evidenced by the patterns of investment flow-ing into the different circuits of capital. Although the deregulation of the finan-cial markets in real estate and the development of new forms of securitization contributed to the rising fortunes of financial firms, their increased riches were also connected to their ability to earn more revenue because of a more gen-eral switching of capital investment out of the primary circuit and into the sec-ondary circuit—a consequence, according to Harvey, of investors seeking out higher returns in finance. While many commentators have attempted to amend or challenge Harvey's rather formulaic account of capital switching by giving more significance to finance as an independent sphere of investment, his argu-ment continues to influence scholars attempting to understand contemporary processes affecting global urbanization (Christophers 2011; Beauregard 1991).

Knowledge as Real Estate:
The Tertiary Circuit of Capital and Fixing Science

In 1983, *Science* magazine published an editorial titled "Knowledge as Real Es-tate." Author Anne Keatley, then senior executive staff officer at the National Resource Council (later the U.S. State Department's deputy assistant secretary for Science, Technology, and Health), argued that science and technology were "highly valued resources" because of their importance for national armament and economic competitiveness (Keatley 1983). But her main contention was that scientists needed to be more cognizant of this value and to become more in-volved in politics in order to have a stronger position and get more resources from the government. What does this have to do with real estate? As far as the casual reader might gather, not much, other than perhaps to suggest that science is the property of scientists. But the article's title points to the deeper relation-ship between science and the new space-time geographies resulting from larger transformations in the economy.

Science is significant for technological advances, and modern industries de-veloped only through concerted efforts in research and development (R & D). In particular, technical improvements in the capacities of machinery and infra-structure enhanced productivity and sped up the production and circulation of commodities. As a consequence, there has been a general decrease in the socially necessary turnover time of capital—that is, "the average time taken to turn over a given quantity of capital within a particular sector, under the normal conditions of production and circulation prevalent at that time" (Harvey 1999, 186).

But firms are still compelled to invest greater and greater portions of their revenues into the development of new technologies—that is, more sophisticated machines but also transportation and communication infrastructure—that help them beat the socially average circulation time. By doing so, they gain a little more revenue than the socially average rate of profit for that sector. These excess revenues are a form of economic rent, sometimes called "technological rents" because the earnings come from productivity increases derived from the exclusive use of a superior technology (Mandel 1975). However, all types of improvements (such as in land or organization) or other extraeconomic factors that indirectly increase productivity are also forms of rent: firms are able to gain revenue above the socially average rate only because of payments derived from unique control of a resource. In fact, as Kraetke has argued, seeking different types of rents is the primary reason firms (often supported by state policy) make significant R & D investments that are designed to produce innovative products and processes (Kraetke 2012).

All investments that firms make into R & D are fraught with different, often contradictory, temporal transformations because in many cases, speeding up in one part of the industrial cycle results in the countervailing slowing down in another part (Schoenberger 1997). For instance, to speed up the production and circulation process and decrease "socially necessary turnover time" requires spatially and temporally immobilizing tremendous investment in new physical infrastructures, which can be recovered only slowly over time. Moreover, buying technically sophisticated labor power can be obtained only at a huge expense, and to recuperate this investment, firms may have to extend the life expectancy of their commodities. The problem is that the relentless push to find new time-saving products and process innovations, which help firms gain revenue above the socially average profit rate for their sector, poses a significant problem—firms need to realize the value locked up in their investments and retrieve its full worth before they are devalued by another wave of innovations adopted by their competitors seeking the same short-term advantages.

In the model of the expanding reproduction of capital mentioned earlier, Harvey suggests that instead of investing into the primary circuit of production, firms could channel revenues into the secondary circuit of physical infrastructure, which includes technically advanced machinery, to indirectly increase productivity. But there is also a third circuit—social infrastructure. Harvey defines the *tertiary circuit* as "investment in science and technology (the purpose of which is to harness science to production and thereby to contribute to the processes which continuously revolutionize the productive forces in society) [or] a wide range of social expenditures [education, health, welfare, ideology, policy, military, etc.] which relate primarily to the processes of the reproduction of labour power" (Harvey 1989b, 65–66). For Harvey, the flow of capital into the tertiary circuit, like in the secondary circuit, can also indirectly influence the processes of production and consumption. However, unlike the secondary circuit, which was supported by the development of financial intermediaries, Harvey holds that the tertiary circuit is mediated primarily through the expenditures of the state (whether national, provincial, or local) and its systems of taxation. On the one hand, investing in science and technology as sources of innovation helps create intermediate inputs that can enhance productivity in the labor process, and, for reasons discussed later, this investment often receives substantial state support. On the other hand, social expenditures, like public schools and subsidies for health care, support qualitative improvements of labor power and the "cooptation, integration, and repression of the labor force" (Harvey 1989b, 66).

Harvey's discussion of the tertiary circuit (as a channel for investment into science and technology) has received little discussion or elaboration, even by Harvey himself. Considering how much his discussion of the secondary circuit of capital and capital switching has revealed about the contemporary urban process since the 1970s, the lack of attention to the tertiary sector is odd; after all, might the growing significance of science and technology for urban competitiveness be related to capital switching?

The most systematic treatment of Harvey's ideas about the tertiary circuit appears in the writings of Richard Walker. Walker, along with Andrew Sayer, tried to account for the contemporary explosion of service-economy employment by arguing that this growth reflected recent changes in the division and organization of labor "along novel lines" (Sayer and Walker 1992, 99). Although Walker does not use the term "tertiary circuit," he draws heavily on Harvey's treatment of science and technology and social infrastructure. Like Harvey, Walker treats a large portion of the tertiary circuit, especially investments in health, education, and civil service, as forms of social consumption that are mediated by the expenditures of the state and designed to increase the overall productivity of labor power.

The difference between Harvey and Walker is the latter's focus on the growing significance of what he calls "indirectly productive labor" in industrial processes (Walker 1985). For Walker, the growth in service-economy employment, and the economy's unique technical and social division of labor, merely reflects indirect labor's increased importance in industrial production. The products of labor can be divided into goods and labor services, and while both have material outcomes and the distinction can be hard to make, only goods are discrete and alienable. According to Walker, the expanding role of indirect labor results from the ways the labor process is being extended in both time and space; how indirect labor is much more hierarchically organized, coordinated, and directed;

and how the mental division of labor is increasingly significant for the social division of labor (certain jobs require great amounts of specialized knowledge, particularly those in the pre- and postproduction processes). Indirect labor tends not to have discrete outputs but rather, Walker contends, to generate information on goods or labor services that enhance the productivity of other parts of a production process and may help firms beat the socially average profit rate for their sector.

For Walker and Sayer, the growing significance of knowledge for industrial production is related to the new social division of labor (Sayer and Walker 1992, 100–101). In most cases, more technically sophisticated kinds of knowledge are increasingly applied directly to producing new products or processes—that is, applied research and development—primarily in specialized preproduction labor. But they observe in a footnote:

> General scientific research is harder to classify than ordinary R&D. It is a highly socialized activity that rarely pertains directly to only one product or labor process, and must be adapted and enhanced for practical industrial use. Most scientific research enters into the stream of available knowledge and cannot be easily commodified, yet basic knowledge about physical systems underlies specific productive activity—much as urban infrastructure lays the basis for city growth. In some cases, however, the products of scientific research can be sold directly as marketable commodities (patents or licensing rights) where they may be classified as information goods or labor services. (Sayer and Walker 1992, 69)

Over the last thirty years, in fact, the revenues from goods that are the products of general scientific research have become much more significant, in part, because of their indirect impact on the production process.

Moreover, Walker and Harvey are keenly aware that contradictory spatial and temporal dynamics are associated with investments into the tertiary circuit. For instance, the technical skills that support the new mental division of labor can be supported only with huge social and individual investment, often at significant state expense—though increasingly, at least in the United States, private student debt is climbing and university research is funded by private sources. Moreover, large sums of money must be dedicated to making new discoveries and developing new technologies, both social and mechanical, that may enhance productivity, but sunken costs are enormous because it is expensive to recruit and pay for the services of highly skilled professionals engaged in research and development.

Like Harvey, Walker assumes that the state is the primary intermediary that supports the production of scientific research and development. Certainly, in the United States, state governments and especially the federal government have a significant role in funding research and development, particularly because of their support for military-related research performed at institutions of higher learning. According to Atkinson and Blanpied, the federal government in 1990 funded approximately 60 percent (down from almost 70 percent in 1975) of all university research (Atkinson and Blanpied 2008, 40). State governments modestly increased their funding for university-related R & D in the late 1970s, but since then, according to Feller, their contribution has remained relatively flat at about 7 percent (Feller 2004, 143).

However, the characterization of the tertiary circuit as mediated primarily by state expenditures may be increasingly less applicable. Despite the government's role in funding the lion's share of R & D, trends since the 1970s suggest that the production of knowledge is becoming more privatized and less reliant on the state. While total spending, both public and private, on R & D has grown over the last forty years, the percentage of the federal government's contribution has fallen. According to Atkinson and Blanpied, in 1975 "the federal government accounted for approximately 45% of total national R&D expenditures, while industry accounted for approximately 42%. . . . By 2000, federal contributions had declined to approximately 24% of the total, with industry contributing close to 70%" (Atkinson and Blanpied 2008, 41). Industry, therefore, was dedicating a larger amount of funding for R & D, particularly for applied research done in-house by private firms (Jaffe 1996, 12,658). Moreover, industry contributions to university R & D programs have risen since the 1980s (Litan, Mitchell, and Reedy 2008, 36). In 1975 industry contributed about 3 percent to universities' research budgets; by 1990 that number had risen to about 7 percent, a level it has maintained for more than two decades. Although this funding remains a small portion of university budgets, these industry-university partnerships directly engage the universities as industrial resources, seek to speed up the transfer of potentially useful knowledge to the private sector, and usually are undertaken to help a specific firm develop new technologies that have immediate commercial benefits.

Nevertheless, the most important development over the last thirty years has been the growing number of universities trying to profit from their intellectual property holdings. While universities have been patenting and licensing their discoveries for decades, since the early 1980s more and more universities have been attempting to derive a growing portion of their revenue directly from their legal ability to alienate the findings made in their research and development programs—even those funded by public expenditure (Mowery and Rosenberg 1993, 18–184). In some cases, universities license their patentable discoveries to a specific firm for a fee, and in other cases they take an equity stake in the firms that commercialize these findings. Regardless of the arrangements, there has been an explosion over the last three decades in the number of university-filed patents and university-supported spin-off firms that commercialize research discoveries made in university labs. A dramatic illustration comes in the filing of university patents: in 1980, U.S. universities filed fewer than four hundred patents, but by 2007 this figure had jumped to over ten thousand. Moreover, the yearly creation of university spin-offs rose from 59 at 98 universities in 1991 to 502 at 155 universities in 2007 (Astebro and Bazzazian 2011). This trend was mirrored in the swelling of technology transfer offices at universities in the United States from around 20 in 1980 to over 140 by 2004—a number equal to that of all the research universities in the United States (Feldman and Breznitz 2009, 166–169). Many authors note that royalties from these patents remain a small portion of university operating budgets (according to Nelsen they represent between 1 and 5 percent of revenues), and only a handful of universities reap significant financial rewards (Jaffe et al. 2007; Nelsen 2005). However, this criticism misses the point that an ever greater number of universities are acting like for-profit businesses and hoping to earn ever larger returns by commercializing their new discoveries.

In the past, the federal government played a special role in R & D funding, especially for research universities, for three basic reasons: (1) there is always uncertainty that untested and new technologies will be financially viable; (2) the commodity of scientific research—that is, knowledge—is "nonrivalrous," meaning that use by one firm does not stop its adoption by a competitor, and keeping it a secret is hard or impossible; and (3) research is information that can be easily stolen or copied at little cost (Astebro and Bazzazian 2011, 256; Hall 2002, 2). Thus, "normal free market" mechanisms failed to support investment into R & D, so government intervention was required. In particular, the contemporary copyright system has been essential to address these market failures by providing legal restrictions on use—that is, limits on the copying of information. Nevertheless, for these three reasons, public entities, primarily universities and federally funded research labs, have engaged in the bulk of R & D that has no proven commercial value, which is often called basic research. In fact, despite a growing financial commitment from private industry toward funding scientific research during the 1980s, in 2004 institutions of higher education still performed more than half of all basic research in the United States. (If other public sources of federal government–funded research were included, the figure would be closer to two-thirds.) Moreover, the percentage of R & D that is performed at universities has actually risen since 1975 because there has been a movement of resources away from federally supported labs toward those housed in universities (Jaffe 1996; Litan, Mitchell, and Reedy 2008).

Although universities continue to perform the majority of R & D in the United States, there has still been growing privatization of the fruits derived from their discoveries. As mentioned earlier, since the late 1970s, universities, including those that are publically supported, have increasingly exercised their patenting authority and sought to reap financial gains from these intellectual monopolies. So despite substantial public financial support, the law ensures that the scientific knowledge produced in universities and alienated into products or processes is not public property but rather part of a university's private financial assets, as are any monetary benefits that derive from that knowledge. Therefore, if new avenues open up that allow a university to profitably commercialize its research discoveries, like in any other private firm, the financial returns flow back into the university's private coffers. Certainly the scientists who conducted the research are financially compensated if their discoveries are successfully commercialized, but so too are university administrators, who look more and more like managers of private firms (Greenberg 2008; Slaughter and Rhoades 2004).

The emergence and expansion of venture capital firms since the 1970s has led to a noticeable financialization of R & D and helped universities commercialize their discoveries. In the 1950s, a small number of venture capital firms played a critical role in regional development of Silicon Valley and Boston's Route 128 Corridor by helping fund new start-up businesses, especially in promising high-technology sectors of the economy that did not have a proven record of profitability; by the 1980s, venture capital funding had greatly expanded and become a central part of the national innovation system (Mowery and Rosenberg 1993). These venture capitalists had realized that there were potentially large monetary rewards in filling a financing gap: while new discoveries were often made in corporate or university laboratories, for whatever reason, these institutions were either unwilling or unable to profitably commercialize their findings (Florida and Kenney 1988). Because of an increasing number of investment opportunities (especially in biotechnology) and changes in federal regulations and taxation, venture capital financing grew rapidly in the 1980s, and major private investment firms, pension funds, and major corporations became significant holders of venture capital assets (Berman 2011). In addition, many state governments and universities established public venture capital funds to support the development of new start-up businesses and, in many cases, sought to capture a profitable revenue stream from the commercialization of research that had been supported by the public purse (Youtie and Shapira 2008; Leicht and Jenkins 1998). While some venture capital firms have helped finance a small amount of basic research efforts, the most significant effect of the venture capital revolution has been to accelerate the commercialization of scientific discoveries and increasingly to make the potential outcomes of basic research (the function that had been assumed by research universities) into a robust site for profiteering.

Productive investments into the tertiary circuit (as productive investments in science and technology) share a number of qualities with investments into the secondary circuit or productive investments in the built environment. The profitable results from investment into science and technology are realized only over long periods of time and at great sunken costs. Insofar as this is true, like investing in the improvement of land, the increased flows of capital into the tertiary circuit may reflect a form of "capital switching" because, like in real estate, investing in science and technology may be a time-extension strategy: the investments are made in the present but are realized only in the far future, if ever. Moreover, investment in science and technology has the capacity to absorb large amounts of accumulated capital that cannot be as productively (as profitably) invested into other circuits. Therefore, while the growing flows of capital into science and technology may appear to be a consequence of a better environment for investment (supported by government policy reforms), these trends may reflect, like in real estate, the incapacity of other economic sectors to soak up large surpluses of capital.

Furthermore, like investments in the secondary circuit, pursuing revenues from knowledge is always a rent-seeking activity, and the revenues that flow from the production of knowledge products or processes, like in real estate, are already a form of finance because they operate as a kind of loaned capital or an advance on an investment that will pay out at some time in the future. In many cases, R & D investments are made because of the indirect benefits they are expected to have for firms by increasing productivity. Here there is a technological rent, as a firm adopting a new technology gains a short-term advantage over their competitors and access to revenue streams above the average profit rate for its sector. However, in other cases the revenue comes from *knowledge-rents*— that is, the anticipated rents (royalties) that may be commanded, even from generalized use, at some future time from the sale of a patent or from licensing revenue. Importantly, when knowledge-rents are pursued, the revenue stream depends not on indirect gains in productivity (although firms are willing to pay a premium for new sources of knowledge to improve productivity) but instead on the ability of the owner of the knowledge to receive a payment for maintaining a monopoly over a unique set of property holdings, even if the knowledge has been universally adopted. The expansion of knowledge-rent seeking has been made possible by intellectual property laws and their ability to make knowledge products into rivalrous goods. Certainly some firms, particularly smaller start-up companies with high concentrations of technically skilled labor, have been able to earn ever greater revenues from the knowledge-rents they command; the same is true for universities. However, this ability to capture rents, and the related investment of financial firms specializing in venture capital, arose not only because of legal developments but also because of the extension of the industrial labor process and the increasing importance of indirect labor inputs.

Conclusion

Geographer David Rigby has noted that theories of regional development in the 1990s were dominated by ideas about the presence of high-technology infrastructure—workers, industrial firms, finance, and so on—but increasingly, the leading explanations of regional development have focused on social relations (Rigby 2000, 216). Irrespective of the particulars, Rigby's point is that economic geographers emphasize extraeconomic factors in accounting for regional growth and competitiveness. Certainly, it seems irrefutable that local conditions influence uneven development and, in part, explain why some regions grow and others do not. Yet Rigby, like many other economic geographers, tends to downplay political factors such as the critical role of federal spending and changes in the national legal infrastructure, as well as the significance of local and regional growth coalitions in mobilizing local, regional, and national resources. Undoubtedly, these factors have to be added to the list of ingredients that shape the uneven geographies of development.

Nevertheless, what I have been suggesting is that both of these accounts fail to address how geographical patterns of industrialization and urbanization in the second half of the twentieth century also reflected the flows of investment into different circuits of capital. While many scholars have focused on the flows of investment into the secondary circuit (the built environment), I contend that the spatiotemporal dynamics of the tertiary circuit (social infrastructure) are vitally important in understanding uneven patterns of growth and competitiveness. Some places, especially those with a large university or a complex of universities, have been the primary beneficiaries of the increasing streams of capital into the tertiary circuit. While Harvey and Walker conceptualized the tertiary circuit as mediated by state expenditure (and it largely still is), there has been an increasing flow of investment into activities that rely on a profitable return from the commercialization of research discoveries. On the one hand, this has been made possible by the extension of the industrial labor process (that is, university research can be more seamlessly embedded into other industrial circuits of production). On the other hand, the development of a legal infrastructure to support the system of knowledge-rent taking has allowed and reinforced the ability to harness the economic value of these discoveries.

In reflecting on the development of a city like Austin since the 1950s, it is hard not to see these large macro processes at work. Certainly, as other chapters show, a significant social and physical high-technology infrastructure developed after the 1970s, largely because of the University of Texas. Moreover, a strong growth coalition led by key actors in the local business community took advantage of local factors as well as federal spending and policy to grow the region, primarily by promoting the expansion of the university's role in the local economy as an employer, a producer of skilled workers, knowledge, and products, and a land developer. Furthermore, the University of Texas has been a leader in commercializing its research discoveries, connecting its patenting ability to regional growth and attracting venture capital financing (Smilor, Gibson, and Dietrich 1990; Ovetz 1996).

Stressing the spatiotemporal dynamics of the flows of capital into various circuits, however, does not preclude attention to local factors, which certainly influenced Austin's ascent in the global economy, nor the specific political configurations and dynamics that have shaped the city. There are a number of unique aspects to the city's development that cannot be accounted for only by macro theories, such as Austin's political dynamics, historical development, and environmental conditions, which the other chapters in this book explore. Moreover, the institutional arrangements and actions of local agents influenced the qualities and features of Austin's distinctive path of urban development. What the attention to local actors also suggests is that other forms of development may also have been possible. In fact, some actors did attempt to create a different kind of city, but these possibilities became foreclosed by a combination of local conditions and unrelenting forces of global capitalism that circumscribe the repertoires of appropriate actions available to cities and their citizens. Understanding how these forces work may open up new opportunities and other possibilities for those imagining or advocating for different forms of regional growth that are more invested in the quality of development rather than its quantity.

CHAPTER 2

The Value of Knowledge

The Expansion of the University of Texas, Urban Renewal, and the Blackland

> The Main University's research program is broader in scope, and more sophisticated than that of any other school in the Southwest. Last year, it received some $6 million in government research grants. This fact has been a boon to industry in Austin. Modern industry wants to be close to high-rated experts with brilliant minds. Over the nation they are building their industries close to universities with highly educated personnel which they can use as consultants and researchists. With this industry, of course, comes greater economic prosperity to the entire community.
>
> **Austin Chamber of Commerce, "Great Academic Stature at UT Blueprinted for 1970"**

> The question cannot be if but how the University is going to expand.
>
> **Austin mayor Lester Palmer**

In 1963 Kerr Clark, a former president of the University of California system, argued that American universities were going through a "second great transformation" (Clark 2001, 65). While research universities could be found in the nineteenth century, since 1950 they dramatically expanded and fundamentally changed their "research role" to become much "larger, more complex, more segmented organizations" (Geiger 1993, viii). Collectively America's research universities became responsible for educating millions of undergraduate and graduate students every year, helping create one of the world's largest highly skilled labor forces (United States Department of Education 2012). Moreover, and perhaps as importantly, they became massive centers for research and development and took on a greatly enhanced role as sites for the production of new knowledge; they became, during this time, the "core of the country's science and technology system" (Atkinson and Blanpied 2008, 46).

By the 1960s it was also apparent that research universities were beginning to have an ever greater role in regional and local economic development. Clark suggested that "what railroads did for the second half of the last century and the automobile did for the first half of this century may be done for the second half of this century by the knowledge industry: that is, to serve as the focal point for national growth. And the university is at the center of the knowledge process" (Clark 2001, 66).

Clark's prognostication turned out to be astute. Knowledge production did become much more central to economic growth, as information and knowledge-intensive activities increasingly formed the leading sectors of the economy because of their potential impact on industrial innovation, and indeed universities found themselves in a new, elevated role (Storper and Scott 2009). University administrators began to encourage, or were persuaded to allow, university professors, particularly in material and natural sciences, to develop partnerships with local firms or establish businesses (Burger 1979; Geiger 1993, 206–309). While these partnerships often happened in research parks developed by universities, it was not uncommon for faculty to become consultants to private industry or to form their own companies (Feldman and Desrochers 2003, 16–20; Luger and Goldstein 1991, 76–154; Lawton-Smith 2006, 128–129). In some cities, most notably Boston and Palo Alto, strong university-industry relations were built, spawning further industrial development either by encouraging new firms to locate to the area or by supporting technological transfer from the university to private firms (Etzkowitz and Dzisah 2008). Moreover, industries, particularly those that were developing in high technology, benefited from the federal defense spending being funneled through universities. It is not a coincidence that the regions of the United States that were the largest recipients of defense dollars, the West and the South (what Ann Markusen called the Gunbelt), were also the regions that experienced the most rapid growth in high-technology sectors (O'Mara 2005; Markusen 1991; Saxenian 1996).

Since the late 1960s, Austin, Texas, has benefited greatly from the presence of the flagship campus of the University of Texas, which has played a major role, directly and indirectly, in supporting the city's development into a high-technology center (Oden 1997; Gibson and Smilor 1991). UT assisted in this development in several ways before 1975. First, Tracor, a defense-related company and the first major local high-technology firm, was set up by UT professors and recruited its workforce and management from the university (Busch 2013). The Balcones Research Center, UT's first research park, was a significant recipient of federal defense dollars for research and was important in supporting a pool of skilled laborers that proved useful in courting companies to locate the first branch plants in the city beginning in the late 1960s (Busch 2011, 141–158). Finally, many UT faculty members, particularly from the faculty of engineering, became consultants for companies with branch offices in the city or in some cases even established their own firms in Austin (Dugger 1974, 159–168).

Yet the significance of the federal government's urban policy regarding UT's development as a major research university and center for science and engineering is hardly mentioned in the scholarly literature on the city's development. Certainly, the federal government, as it was for all research universities in the United States, was the central source for research funding at UT, especially defense-related research. But more importantly, what is missing from the scholarship is that for UT to become a more robust research university in the 1960s, like many other universities throughout the country in denser urban areas, federal urban policy was required to assist in its land-acquisition program. Today, more and more scholars are focusing on the "spatial consequence of growth for higher education," especially how the federal government's funding for land development facilitated the significant spatial expansion of research universities in the 1960s (Winling 2010, 2). What can be seen in UT's development into a much larger, more complex research university in the 1960s and 1970s is how much it was connected to both increases in federal research funds and the federal urban policy that helped make possible the expansion of the main campus.

In this case study of the University of Texas at Austin, I show how the motivations behind a substantial land-acquisition program undertaken in the 1960s were different from those of previous territorial expansions and were a central part of the university's transforming role as a research university. I also describe how the alliance among national and local actors reshaped urban policy and influenced the production of certain economic practices and the land-use possibilities of the university. Implemented in a particular way by the City of Austin and its agent the Austin Urban Renewal Authority, supported by the state legislature, and adopted by the Board of Regents of UT, federal urban renewal policy was especially enabling because it helped establish the legal framework for taking land and deferred the financial and social costs associated with the university's expansion. However, it is important to point out that the social costs of the expansion of the main campus were distributed unevenly in space and disproportionately borne by certain groups of people, namely more modestly resourced communities of Hispanics, Caucasians, and particularly African Americans.

The university's capacity to remake Austin's urban space, and the subsequent impact on the regional economy, draws attention back to the fundamental role that federal urban policy had in the production of the knowledge economy. For decades the scholarship on Austin has either focused on the local growth coalition or made too much of the university's positive "spillover" in developing the region. Like other more contemporary scholarship, I argue for a new avenue of academic inquiry that sees UT as having a central role in the economic development of Austin into a high-technology region because of its role as a land developer, a role that was assisted by federal legislation.

Federal Urban Policy and the Changing Mission of the Research University

Beginning in the 1950s, after the passage of the Housing Act in 1949, the federal government began to take on a much more active role in urban policy. In particular, the term "urban renewal" emerged to refer to the federal program that provided funds to municipalities mainly for slum clearance made possible by Title 1 of the 1949 act. Under its new authority, the federal government could provide funds to a local government (or its agent, a development authority) for the costs of purchasing existing homes or parcels of land in an area and razing, rebuilding, and/or rehabilitating them. The law also authorized federal authorities to provide financial assistance to relocate or rehouse residents in the affected communities. Expanded several times over more than a decade, the program grew in scope and increasingly allowed for federal funds to be used for an ever greater number of programs, particularly commercial developments, while it decreased its involvement in housing projects (Teaford 2000).

Urban renewal reshaped the landscape of cities, in particular their racial geographies. Established because of the demands made from various constituencies to stop the decline of older neighborhoods in cities (particularly near their central business districts), urban renewal efforts involving land clearance and redevelopment disproportionately affected urban neighborhoods with the highest concentrations of African Americans. While federal policy was said to be racially neutral, efforts by local governments were so brazenly racialized that in a 1967 interview, James Baldwin, the famed African American novelist, said, "Urban renewal . . . means moving the Negroes out. It means Negro removal; that is what it means. The federal government is an accomplice to this fact" (Baldwin 1963). In fact, for about three decades of the twentieth century one can find many cases where local governments, under the aegis of federal policy, claimed they were improving their cities by clearing blighted neighborhoods and then building new, often public, housing for the affected communities. In reality, however, African Americans were dispossessed of their property; their communities were torn apart, and in many cases the residents were just moved from one degraded neighborhood to another (Teaford 2000).

One thing that became more apparent over time was that when land clearing took place, especially in areas where a commercial redevelopment would be desirable near city centers, urban renewal "was not an anti-poverty program [but] an effort to inject economic vitality into central cities" (O'Mara 2005, 76). Moreover, local progrowth urban coalitions emerged that used this federal policy to transform cities, particularly their older central business districts, into "postindustrial administrative centers" (Mollenkopf 1983, 140). So while the federal government did collaborate in these land clearances, which provided local economies with new rounds of profitable investment in urban centers, their intervention was more than just an urban clearance program: the government's urban policy, in many cases, was an economic policy because it supported the reorganization of urban space to facilitate the development of particular kinds of economic practices, which would now be classified as the "postindustrial knowledge economy."

As a result of lobbying efforts, various universities' expansion plans became conjoined with federal urban renewal policy under "Section 112" in the amended 1959 Housing Act and became central to the remaking of cities for the "new post-industrial age" (O'Mara 2005, 79; Winling 2011). For example, William Slayton, then commissioner of the federal urban renewal administration, wrote in a 1963 report:

> For the past several years there has been a remarkable growth in the interest and participation of universities in urban renewal. Part of this interest is an outgrowth of the great pressure on universities to expand their physical facilities to meet rising enrollments. During the past 30 years enrollments in colleges and universities have more than tripled, and are expected to double or triple again during the next 20 years. The task of preparing for this growth is overwhelming, and must fall largely upon existing institutions. But new facilities cannot be created unless there is land available, and most of our urban universities are landlocked. While they have been making major efforts to acquire land on the open market, this has proven very difficult. The universities have come to realize that urban renewal can assemble land needed for expansion, when the land is blighted. Accordingly they have become active redevelopers. . . . The university interest in urban renewal is matched by that of municipal officials who understand the important role that universities play in the economic and cultural life of the community. The universities make a significant contribution to the local economy with their large payrolls and the substantial expenditures by their students. But even beyond these direct expenditures, the presence of a university and its technological and research resources serves as a major factor in attracting industry to the community. The Congress recognized the importance of the university-municipality relationship and the role urban renewal could play in enhancing this relationship when it added section 112 to title I in the Housing Act of 1959. Section 112 offers two advantages to renewal projects which provide space for university expansion or create a more compatible environment for university functions: (1) Such projects are exempt from the residential requirements of title I and (2) certain expenditures by the university in acquiring land near urban renewal projects may be credited toward the local share of carrying out the project. (United States Congress 1964, 429–430)

Under these federal laws, institutions of higher learning across the country expanded their campuses with federal assistance (Parsons and Davis 1971). Moreover, the provisions in Section 112 made incorporating universities into urban-renewal projects more attractive because of the two-to-one credit program now offered. According to Winling, "University contributions near to and consistent with an approved urban renewal plan would trigger a two-to-one federal match, even for acquisition or demolition expenses up to five years before the enactment of the renewal plan. Federal funds in excess of the cost of the project could become credits transferable to urban renewal projects anywhere in the city" (Winling 2011, 70). While some accounts of university expansion chalked up these efforts, perhaps in part correctly, to the imperial aspirations of administrators thirsty for more land and students (that is, having bigger campuses), it is important to stress, as the quote from Slayton reflects, to what extent the research demands of universities and their growing economic value cannot be divorced from the spatial policies that supported both.

Furthermore, the federal government's expanded role, as a primary financial supporter of university research, is part and parcel of university land expansion plans. In 1953 the federal government's contribution to funds available for research at universities was about 55 percent; by 1970 this had risen to almost 70 percent. Most of these funds went to research and development with military-related purposes, though some non-military-related research, especially in the natural sciences, education, and health, also benefited (Graham and Diamond 1997, 47–48). Moreover, as Geiger has shown, the federal government's share of research funds to the top twenty-four research universities (the universities that received the bulk of the funds for research until 1960) decreased as the federal government funded an ever broader number of universities. As a consequence of these new opportunities for funds, an intense interuniversity competition ensued as more and more universities saw a chance to emulate "more prestigious models" and rise in the academic hierarchy (Geiger 1993, 203–204). Position in the hierarchy was essential because it, along with political connections, ensured the ability to garner more and more federal funds (O'Mara 2005, 52–55). One way to improve your status was to establish well-funded graduate programs. Another way was to improve the quality of the faculty by hiring more researchers who were well regarded in their respective fields. Both of these tactics could be accomplished only by making more money available for research, but they also required spending more money on facilities such as labs, buildings, and offices.

Local progrowth coalitions sought to capitalize on the "need" for universities to expand, which was created by this new federal role in research, and remake their cities into what O'Mara has cogently called "cities of knowledge." A new kind of urban space could be produced, one that was dominated and controlled by a university, and whose primary purpose was the production of knowledge. Increasingly, therefore, universities saw themselves as partners with local growth coalitions in land-development programs and as potential partners with industry in promoting technological innovation and regional growth. Moreover, universities began to take on a new role as urban developers.

However, as universities increasingly became involved in land development, they more and more became the target of animosity from their surrounding communities. For instance, Stanford, in its quest to propel the development of technology in Silicon Valley, encountered substantial opposition from local environmentalists (O'Mara 2005, 132–139). In existing urbanized areas, such as

Philadelphia, Cleveland, Detroit, Baltimore, Newark, and Chicago, university expansion efforts, supported by federal and state urban-renewal laws, met fierce resistance because these plans often required the removal of people with more modest resources, primarily African Americans, from their neighborhoods (Hirsch 1998; Ernst et al. 1974; Winling 2011; Parsons and Davis 1971; Souther 2011). Such a reappraisal of university expansions, and their devastating consequences for some communities, is increasingly forcing reconsideration of the sometimes-Pollyannaish account of postwar high-technology growth in cities, an account that seems particularly strong in Austin.

The Making of the University of Texas and the 1921 Proposal to Move the Main Campus

The University of Texas system is a quasi-state institution. Since 1881, an independent Board of Regents, appointed solely by the governor, has governed the system. Originally, the UT system consisted of only the main campus in Austin and the medical school in Galveston, but it now encompasses nine academic campuses and six medical or science schools across the state. The system's chancellor, vice-chancellors, and all the university and school presidents that are appointed serve at the pleasure of the regents (Dugger 1974, xvii). However, the university is considered an appendage of the state, as is its property. Therefore, although the university earns huge sums of money from mineral royalties, cattle-grazing rents, and tuition, it does not pay federal income taxes, nor does it pay land taxes to any of the municipalities where it administers a campus. Moreover, as a state institution the university was always able to take land at a fair market price with authorization by the State Legislature, but since 1965 the power of eminent domain—the right of the government or its agent to take private property for public use at a just compensation—has been granted directly to the regents.

One of the most peculiar aspects of UT is that it is supported by a state-created sovereign fund known as the Permanent University Fund (PUF). This financing structure has important consequences for the university's development. Like other universities, it receives operating revenues from tuition, grants, donations, and interest or dividends from its endowment. Moreover, as is typical for public universities, some of these funds come directly from the state government's coffers; presently, that contribution makes up about 15 percent of the university's operating budget. However, since 1876, UT has been financed largely by income derived from its public endowment. The PUF's principal asset is a huge portion of land in the state—over 2 million acres, nearly all in western Texas—granted in several legislative appropriations in the nineteenth century (Erwin 1973). The 1876 statutes that created the PUF stipulated that the revenues that come from control over this land, taken from either mineral (subsurface) royalties or the land's disposal, would form the endowment's principal, and these revenues had to be reinvested into the PUF. In contrast, revenue derived from the "surface," primarily grazing leases, land rents, or dividends paid from investments made from the PUF such as interest-bearing bonds (later amended to include dividends from all financial securities) could be used only for constructing buildings (personnel were primarily paid by state legislative appropriation or tuition) (Cook 1997; Prindle 1982, 278; Matthews 2007, 22–24).

Moreover, UT is one of the largest public universities not created by a federal land grant. In 1876, when the State Legislature established the PUF, it also mandated that land for the first state university "be located [somewhere in the state] by a vote of the people of this State" (State of Texas 1876). Although UT had been given a large land grant for its endowment (and later its holdings were dramatically expanded by further legislative action), it took several more years for construction in Austin to begin, and it was only in 1884 that the first building was completed (Long 1964). In part, the delay was due to the demand for a vote to determine where the campus should be placed. After a competition among several cities throughout Texas, it was decided in a state referendum in 1881 that the main campus of the university system should be built in the newly formed Texas capital in Austin and that a medical school should be constructed in Galveston. That same year the governor signed an act that both set aside forty acres in Austin for the campus and vested the regents with control over the university.

However, quickly it became apparent the available funds created by the PUF were inadequate to meet the demands of expanding or maintaining the university's physical plant. Nearly all the land the legislature had granted into the PUF was considered not very valuable. It was not particularly good for farming or grazing, and little profitable mining had been discovered. Plus, because it was far from large urban settlements that may have inflated the value of its real estate holdings, the land had a minimal resale price. To make matters worse for the university, the legislature in 1876 had also prohibited the use of any state funds for the construction of new buildings. "As a consequence," Prindle writes, "for decades most of the buildings on the campus in Austin were miserable shacks" (Prindle 1982, 278).

Despite the university's paltry conditions and resources, enrollment grew rapidly, and by 1910, only about ten years after the first main building had finally been completed, the spatial limitations of the campus became apparent. The area around the main campus in Austin had been encircled by residential neighborhoods, but as the student population had grown, it was evident that additional facilities needed to be built and existing ones to be upgraded (Brown 1984; Erwin 1973; Holland 2006, 91). After years of internal deliberations, the regents voted in 1921 to move the entire campus to a five-hundred-acre plot in another, less centrally located, section of Austin—a plot donated to UT in 1910 out of concern that the university would not be able to expand at its original location (Long 1964). In their request to the governor and legislature, the regents noted that enrollment had increased more than 100 percent since 1914—almost 20 percent a year—and that there was already insufficient space for boarding and educating these students. Complaining that the buildings were "inadequate, outworn, and dangerous" because many were old, wooden, and not fireproof, the regents argued there was no option but to raze them and sell the land (Long 1964, 91). Houses occupied almost all the land around the university that, the regents claimed, could be legally acquired but could not be taken without the "costs of the bitterness" to the surrounding community, "which such a course must inevitably produce" (Long 1964, 92). Lastly, the regents argued that an expansion at the present location could happen only at "prohibitive" financial cost (Long 1964, 92). What the regents did not say is also significant: nowhere in their statement to the governor is there a mention of the needs of professors, the university's research profile, or the university's scholastic ranking.

Nevertheless, in the end, the main campus of the university did not move from its original location. After fierce opposition from the Chamber of Commerce, local businesses, and powerful former alumni, the state legislature responded by appropriating over $1 million to buy up 135 acres of land adjacent to the university, and the original 40 acres expanded to 175 (Austin American-Statesman 1956; Brown 1984, 31; Long 1964). What proved to be perhaps more important for the upgrading and expansion of the physical plant was the discovery in 1923 of a huge reserve of oil under the lands owned by the university (Vinson 1940). In a mere three years, from 1923 to 1925, the oil royalties in the PUF went from about $16,000 to almost $4 million (Prindle 1982, 286). Needless to say, the size of the endowment greatly expanded, but because oil is "subterranean" it was considered part of the principal of the endowment. Thus, the massive amount of revenue flowing into the fund could only be reinvested into the PUF. Although the net income increased dramatically from interest on investments made with this new revenue, as did the money available for expanding or improving the university's physical infrastructure, these funds were still insufficient to meet the ever growing demands placed on the university. In part to alleviate this pressure, in 1928 the legislature began to allow the university to issue bonds against a portion of the endowment's principal to finance new construction for permanent improvements. (This authority has been expanded dramatically over the years, with significant implications for the university's ability to raise funds for new land development; see chapter 3.) With its new bonding authority and newly acquired land, one of the largest expansions in the university's history was undertaken on the main campus in Austin from 1922 to 1942 (Cook 1997, 20; Speck 2006).

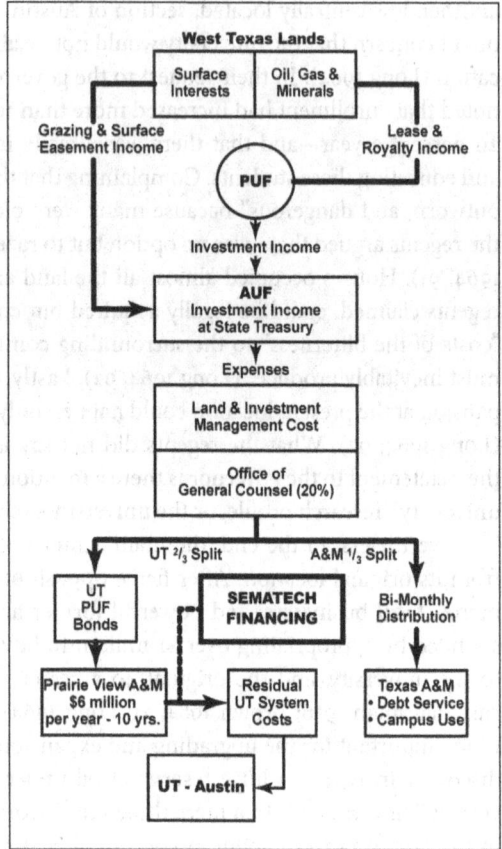

West Texas Lands

Surface Interests | Oil, Gas & Minerals

Grazing & Surface Easement Income

Lease & Royalty Income

PUF

Investment Income

AUF Investment Income at State Treasury

Expenses

Land & Investment Management Cost

Office of General Counsel (20%)

UT ²/₃ Split — A&M ¹/₃ Split

UT PUF Bonds | SEMATECH FINANCING | Bi-Monthly Distribution

Prairie View A&M $6 million per year - 10 yrs. | Residual UT System Costs | Texas A&M • Debt Service • Campus Use

UT - Austin

The Committee of 75 and the Birth of a New University Model

In 1957, on UT's seventy-fifth anniversary, the regents appointed some of Texas's most powerful elites to "undertake a 'realistic and sober reappraisal' of the University of Texas as it enters the final quarter of its first century" (University of Texas System Committee of 75 1958, Preamble). After a series of surveys and interviews, the Committee of 75, as it was known, issued a report that outlined areas where the university should be improved. The state constitutional statute that had created UT as an institution for "public education" demanded that the university be "of first class," but according to the committee, this goal had not been achieved (8). While the undergraduate and graduate programs were given a "satisfactory ranking," the university research program was deemed "less than satisfactory." The problem was that "the research program has not attracted many great scholars . . . and little of the work [had] won national recognition" (9). The committee, therefore, recommended that the "most dramatic single change . . . during the next 25 years should occur in the character and quality of its achievements in the field of research" (15). In order to achieve this goal, the university would have to offer more competitive salaries and a better working environment, which included increasing the research facilities available to faculty (17–18).

One of the central problems in improving the work environment, and therefore the rank of the university, was the condition of buildings and other research facilities (University of Texas System Committee of 75 1958, 20). Physical plant improvements, less hindered by financial constraints than before, were still hindered by spatial constraints. The university was located in the middle of Austin, in a well-built-up area, and there was no readily available land around it for new development. But making better research facilities required, in part, expanding the land occupied by the university. The report read as follows: "The Main University should seek Legislative permission to purchase additional land contiguous to The Main University campus to permit expansion of its facilities in pace with growing needs. It is clear that a major expansion of The Main University will be imperative over the next 25 years, and many difficult and costly problems can be avoided if careful thought is given to future requirements and if steps to acquire land are taken in the proper time" (21).

Significantly, while nearly all of the report from the Committee of 75 was adopted verbatim in the official Board of Regents public response, the above statement was omitted (University of Texas Board of Regents 1960e). However, I suggest that this omission was due not to the regents' lack of endorsement for this idea but instead to the public anger it may have caused had it been officially sanctioned.

Furthermore, I argue, the subsequent expansion plans of the main campus of UT at Austin over the next forty years should be seen in light of the recommendations of the Committee of 75. To support my contention, one need only turn to what the administrators at UT actually did. Approximately two years after the committee made its recommendations, the university undertook a confidential audit to figure out how to implement these recommendations into a master plan. The series of memos drafted for the administration regarding changes to the physical plant addressed student enrollment, faculty needs, sports facilities, classrooms, administrative offices, and so on. The report was clear that "space-demand is somewhat separate from gross enrollment" (University of Texas Board of Regents 1960a, 12). It stressed that "faculty office needs are growing much faster than enrollment" and that a "drastic action to stop enrollment increase could reduce building needs 1961–70 by $8 to $10 million at the outside; quickly enacted freshman admission restrictions, imposed gradually, could result in $4 million less in building" (12). Undergraduate enrollment, in other words, could be limited for the purpose of saving money for faculty research, which was the "valuable means for pursuing eminence" (University of Texas Board of Regents 1960c, 36).

In memo 3, titled "The Problem of Main University Space and Its Utilization," the auditors explored how campus land was being used. Here, recommendations were made based on a "space audit," which took a map of the campus and drew red lines around an area that was estimated to be the largest zone in which a student could walk to class in ten minutes. The findings of this space audit suggested that peripheral (e.g., sports fields) or support (e.g., physical plant) services should be moved out of this inner campus area in order to make room for student and faculty services. However, it did not suggest where to relocate these services, nor does the report seem to endorse acquiring new land near the university: "Several apparently desirable land acquisitions contiguous to the Main Campus have been mentioned. In one sense, it appears rather wasteful to be purchasing land when we have the large acreages of Brackenridge and Balcones [Research Facility]. Limited utility of these areas is increasingly apparent, however" (University of Texas Board of Regents 1960b, 29).

So while the auditors still suggested that auxiliary and support facilities could be located elsewhere on land the university already owned, they recognized the problem associated with their spatial division from the main campus. However, when they discussed the possibility of acquiring more land east of Red River Street, they offered several caveats about why this would not be a good idea.

Although several suggestions have been made for acquisitions of tracts East of Red River, some of which are relatively unimproved, no very pressing need can be identified. In the far-distant future the academic campus could extend along a shuttle-bus route all the way to East Avenue, but this seems a rather remote possibility. Any storage facilities or shops located East of Red River could be put at Balcones with slight additions to operating cost. Football fans would welcome parking lots in this area, but wouldn't pay enough to amortize grading and topping, probably. No one who has studied the use of satellite parking for students at these distances is very sanguine about its acceptability—especially if a fee has to be charged for its use. (University of Texas Board of Regents 1960b, 29)

In other words, the auditors acknowledged that it was be possible for the university to expand eastward but noted that it was not feasible to move academic facilities (the main source of the university's problems) that far to the east. Moreover, they suggested that even relocating support services to an acquired area to the east was not a good idea because while they would be closer to the campus, these services could just as easily and at minimal extra costs be moved further offsite to land already owned by the university. Lastly, parking, a major reason given for the needed expansion, was a notable concern, but it too would not be cost-effective because it would not pay for itself.

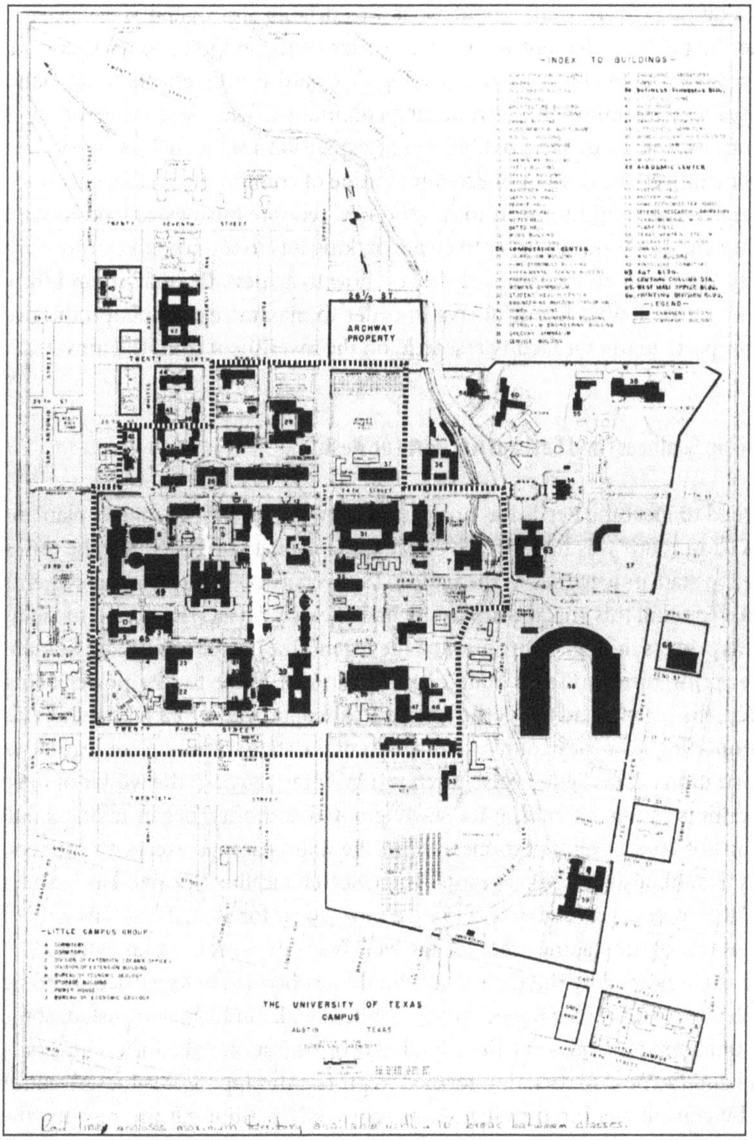

FIGURE 2. The Space-Auditor's Map of UT Austin's Main Campus, 1960.
Board of Regents Real Estate Records, Dolph Briscoe Center for
American History, the University of Texas at Austin.

The auditors were frank about what was driving the spatial shortage, but unlike the public statement mentioned earlier concerning the 1921 expansion, the auditors' statement showed a striking disregard for the effects of any land acquisition program on the surrounding community. There was no mention of community concerns, the possibility that expansion might fuel animosity toward the university, or the noneconomic value of community in the surrounding area. What would it mean to raze homes, relocate businesses, and disrupt several neighborhoods in order to create parking lots to service fans for football games? The auditors' focus was clearly on how to achieve the maximum utility of space at the lowest financial cost in order to maximize possible productive returns, particularly for faculty research, on the investment from the university.

Expanding Southeast and East but Not North or West

Designed to meet the needs for "10 years of growth," the official master plan announced in 1960 "was intended to boost the University of Texas into the ranks of the top state universities in the nation" (University of Texas Board of Regents 1960d). To fulfill this goal the university had to expand its territory, and to facilitate this process the legislature granted the regents the power of eminent domain in the spring of 1965 (Texas State Legislature 1965). Prior to this power being granted, the regents had to ask the legislature for approval before taking individual properties; now they could act as agents of the state without prior legislative authorization (Texas State Legislature 1965; Waldrep 1965). In the winter of 1965, the administration, exercising its newly granted authority, began taking small portions of land in all directions around the main campus except to the west (Daily Texan 1965b; Maddigan 1965). Then vice-chancellor Norman Hackerman said, "it is 'proper-sighted' to gain space now [1965] for 1975. The University of Texas is one of the staging areas for the worldwide explosions of knowledge. . . . [T]he space needed by the University should be obtained before prices become completely prohibitive" (Brewer 1965). While not all landowners resisted, some were unhappy to dispose of their land. An organization called the University Area Land Owners Association formed to protest the appraisal values given for land by consultants hired by UT. After repeatedly petitioning the regents, the legislature, and the courts, these landowners were able to receive greater compensation for their land than they were originally offered (Daily Texan 1965a). While this action was unable to stop the university's land acquisitions, it made the process much more difficult and much more expensive for the university than was anticipated (University of Texas Board of Regents n.d.).

While the main campus had expanded since its founding in 1876, the proposed expansion under the master plan in the 1960s was different. Although not as dramatic as the 1921 proposal to move the entire campus, the master plan had the potential to change the layout of the campus. The *Alcalde*, the alumni

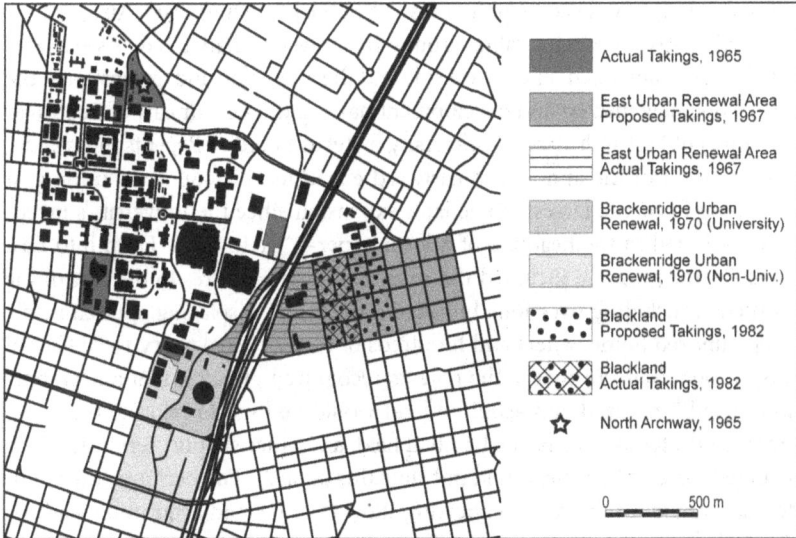

FIGURE 3. UT Austin Land Takings, 1963–1988.

newsletter, featured a whimsical headline that alluded to the removal of students from the university: "Are we going next?" In the 1965 article the major concern was the rumors that the administration, in order to comply with the recommendations of the "space auditors," discussed moving the football stadium, which was located in the center of campus. Very controversial to many alumni, this plan was ultimately abandoned (University of Texas Board of Regents 1965b, 129–130). More telling, however, was the subtitle of the *Alcalde* article, "The Expanding Boundaries of the Forty Acres move North, South, West, and mostly East" (Brewer 1965). But how far the university would try to extend eastward the reporter would not have known.

In 1966, despite the recommendations from their own internal "space-audit," the regents announced that the main campus would undergo a massive eastward expansion on a scale well beyond what anybody could have imagined. Suggesting that it was "impossible to expand in any direction without inconveniencing a considerable number of people," William Heath, chairman of the Board of Regents, argued, "It is completely impractical for reasons we have given many times for the University campus to substantially expand except east" (Heath 1966). Complaints about future spatial limitations and concerns about the need to push the university eastward seemed to have been widely circulated by members of the university community as far back as 1959. "A problem that is not so worrisome is the matter of 'where do we grow' geographically," Jo Eickmann at the *Daily Texan* wrote in a manner reminiscent of the geographer Friedrich Raztel; "The University's room to breathe in lies principally to the east of campus" (Eickmann 1959). Moreover, the necessity of taking a large amount of land to the east may have become more pressing in the preceding years. An appraiser UT hired to assess property values around the university had noted, "As available land has been absorbed on the west side and the relocation and expansion of the law school, new interest and development has begun to occur east of Red River and north of East 19th street" (University of Texas Board of Regents 1965a).

The extent of the announced eastward expansion in 1966 was stunning. It was going to be the largest single land expansion since the university's inception. Reflected in the headline of a local paper—"UT to Knife Deep into East Austin"—the proposal included the seizure of over 140 acres to the east of the university, including over four hundred parcels and displacing approximately three hundred homeowners and hundreds of renters (Castlebury 1966; City of Austin 1966; Ransom 1966). The regents recognized privately that a neighborhood would be razed and acquired land would be used to relocate the intramural fields, tennis courts, the baseball field, and other support services, which included parking lots, but would not state this publicly. Furthermore, the confidential minutes of the regents' meeting show that the land acquisition was primarily for future use (University of Texas Board of Regents 1966b).

Publicly, the rationale for such a large land acquisition was the fear that the university would have to limit student enrollment; after all, enrollment at the main campus had grown from about 18,000 in 1950 to about 25,000 in 1965, and it was projected to rise another 10,000 in just ten years, which it did (City of Austin 1965, 19–20, 1974, 14–15). "If we are not able to acquire additional usable land then we'll have to consider limiting our enrollment," threatened Frank Erwin, chairman of the Board of Regents. "If we limit enrollment, instead of the Main University being the principal university in the state, we'll find others being the largest and the principal schools" (Castlebury 1966). Austin mayor Lester Palmer echoed a similar theme: "A few people may be inconvenienced but on the other hand many hundreds of thousands of young students may not be afforded an opportunity to attend the university" (City of Austin 1966). Yet the position taken by Erwin and Palmer seems less than candid. Certainly the expansion plans took into account concerns about enrollment, but internal memos make clear that faculty research needs, not enrollment, primarily drove the enlargement policies. However, such a justification, stated publicly, may not have been as persuasive because it would have been inconsistent with the widely held belief that educating students should be a public university's primary mission (Dugger 1974).

The Blackland or the Winn Tract?

The question looming in the background, meanwhile, was why an eastward expansion, an expansion that would inconvenience a considerable number of people, seemed more logical than an expansion in any other direction. To be sure, the decision was driven by financial costs; the properties east of the main campus were less expensive and could be acquired at a "cheaper" price. Monetary value, however, masked the social price UT could not afford to pay. While the administration's territorial aspirations may not have had limits, there were social limits that shaded the possible geographical confines of the planned expansion. The proposed land seizure was going to destroy only a neighborhood that was more modestly resourced, and nearly all of the people that were going to be affected were nonwhite minorities, particularly African Americans. Operating under what Abrams called the racist theory of value, UT administrators and city officials worked with the assumption that African American neighborhoods, households, and bodies were simply less valuable and desirable than those of whites. Since the university was going to change the "color" and character of the neighborhood by removing its inhabitants, it was also going to improve its value (Abrams 1955, 157–158). Yes, the land was less expensive, but this was the result of decades of antiblack racism in Austin. Focusing on only the financial price obscured the social relations (unequal treatment of whites over blacks) that had generated and continued to reproduce the devaluing of the area and its peoples.

From the outset, the administration claimed that the proposed taking of land was both necessary and desirable. It was necessary for the university not to limit enrollment and to maintain its position as the premier place for higher education in Texas. Land taking was desirable because it would improve UT's national academic profile and boost the long-term economic fortunes of Austin. Moreover, local officials were strongly in favor of the university's taking the land to the east because it would remove a "poorer," mostly "blighted," neighborhood.

The area to the east, locally known as the "Blackland," had originally been called the "Blacklands" by Swedish farmers who had coveted the area in the nineteenth century for its rich and dark "Blacklands" soil. (The Blackland Belt or Prairie is an ecological subregion in eastern Texas, famous for its agricultural production, especially cotton [Foley 1997, 30–32; Prout 2012, 141–143].) As Austin's population grew, the farms just east of the capitol were subdivided into single-family homes, and by the 1950s the area had become occupied primarily by African Americans. By 1960, the "s" had been removed from the colloquial use of the name, making it the "Blackland," which was sometimes referred to pejoratively as "niggerland" (McCarver 1995, 21).

Prior to the 1930s, African Americans had been located throughout the city in small numbers, but their largest grouping was in an area east of downtown (Tretter 2013). Several factors, discussed in more detail in chapter 6, pushed African Americans to concentrate in neighborhoods in the eastern part of the city, but two issues should be mentioned now. First, the 1928 master plan adopted by the city recommended that all social services such as education and hospitalization be made available to blacks only in the eastern part of the city. Second, racially restrictive covenants, which forbade nonwhites from access to housing, proliferated in Austin's newer western and northern neighborhoods. Over time, the Blacklands and other older neighborhoods, which were located close to other neighborhoods with high concentrations of African Americans and which also lacked restrictive covenants, became the areas of immigration for nonwhites from other areas in the city and elsewhere. As whites moved out or died, the demography of the area gradually changed from being a working-class white, primarily Swedish neighborhood to one dominated by African Americans. As it became an African American neighborhood, the area began to suffer from chronic problems, such as lack of investment and disenfranchisement caused by racial discrimination, that produced declining social and housing conditions.

The legacies and realities of antiblack racism that supported UT's expansion plans are perhaps most clearly seen in how representatives from the university and the city government referred to the area that was going to be seized. Despite the wide acceptance of the neighborhood's name as the Blackland, UT and the City of Austin officials referred to the project area as the Winn Tract (named after the elementary school located in the area), the University East Project, or the Eastern Renewal Area. UT had a poor track record when it came to the treatment of African Americans, and Goldstone contends that during this period regents, sensitive to charges of racism, often tried to distance the university "from the more blatant forms of racial segregation" (Goldstone 2006, ix). Since its founding in the 1880s the university had admitted only whites. In 1950 UT was forced by the Supreme Court to integrate its professional and graduate schools, and while African Americans were able to enter its undergraduate program in the 1960s, its dorms and sports teams remained almost completely segregated until the 1970s. During the 1950s and early 1960s, desegregation protests were common at the university, and the main campus was an important staging ground for antiracist student activists throughout Texas; later, around the time of expansion, UT was "the largest center of new left activism in America" (Kuhlman 1995; Rossinow 1998, 9). Perhaps the regents were fearful of referring to the neighborhood as the Blackland because doing so risked unmasking the social processes that allowed for the entire eastward expansion and enlivening these activists against their proposal.

The Racial Effects of Urban Renewal:
The Brackenridge and Eastern Renewal Areas

The regents had not come up with the idea of taking such a large section of the city. The plan, as confidential and public documents show, was presented to the administration by the officials at the Austin Urban Renewal Agency (AURA), a private, nonprofit, quasi-governmental organization established in 1959 by the Austin City Council for slum clearance and blighted-neighborhood rehabilitation. Made up of all the City Council members plus some additional planners and local stakeholders, the AURA had been vested by the City of Austin as a legally separate nonmunicipal entity with the powers of eminent domain mandated to promote urban renewal, which often meant razing "deteriorated neighborhoods" and selling the land back to private or nonprivate actors for redevelopment (Austin Urban Renewal Agency 1972, 2). Backed by the federal Housing Act and supportive state legislation, the City of Austin, like countless other municipal authorities, was able to give these private entities the legal authority to forcibly buy real property for the purpose of removal or rehabilitation. Once an urban-renewal agency decided that a selected area of a city should have the legal designation "slum," that area could be seized and slated for land clearance, and the federal government would pay for part of the land seizure and the rehousing of the displaced homeowners and renters. The appropriated land would then be sold back to other developers for "higher and better" uses.

Before AURA was formed, the City of Austin, in 1954, created an exploratory committee called the Greater East Austin Development Committee (a committee with almost no representatives from East Austin) to "study the needs of the East Austin Community," and "it was through this committee that the objectives of the Urban Renewal Program were formulated." Those objectives included conservation, upgrading, clearing, and redevelopment (Austin Urban Renewal Agency 1972, 1). In 1957 the Austin Urban Renewal Department proposed a series of urban renewal projects targeting several "areas of concern" that had been suggested for renewal by the Greater East Austin Development Committee. Almost immediately, private property rights advocates sued the city, and the Texas legislature responded by passing the appropriate enabling legislation, the Texas Urban Renewal Law, which permitted the creation of municipal "urban renewal agencies" to clear designated "slums" or rehabilitate "blighted" areas with federal financial support if the agencies were established by local referendum. Two years later, the referendum passed in Austin, and after the defeat of pending lawsuits, city officials proceeded in 1962 to create the AURA.

UT's attempt to seize such a large section of the city at one time was possible only because of the AURA. "Basically we came up with the proposition that the University's expansion to the east was feasible," city manager Bill Williams told reporters, "and could be tied in to [sic] an urban renewal program" (Castlebury 1966). UT's expansion plans seem to have been of particular interest to officials within the AURA. A 1965 map created by the agency marked areas to the west and east as areas of concern and therefore potential targets for urban renewal. Certainly the agency was aware that UT was buying up nearby land as part of an expansion effort and that this effort was disruptive to neighborhoods around the campus. Involving UT in the urban renewal program would have allowed the agency to channel the university's growth patterns. It also relieved AURA of the burden of seeking out developers because the university would be

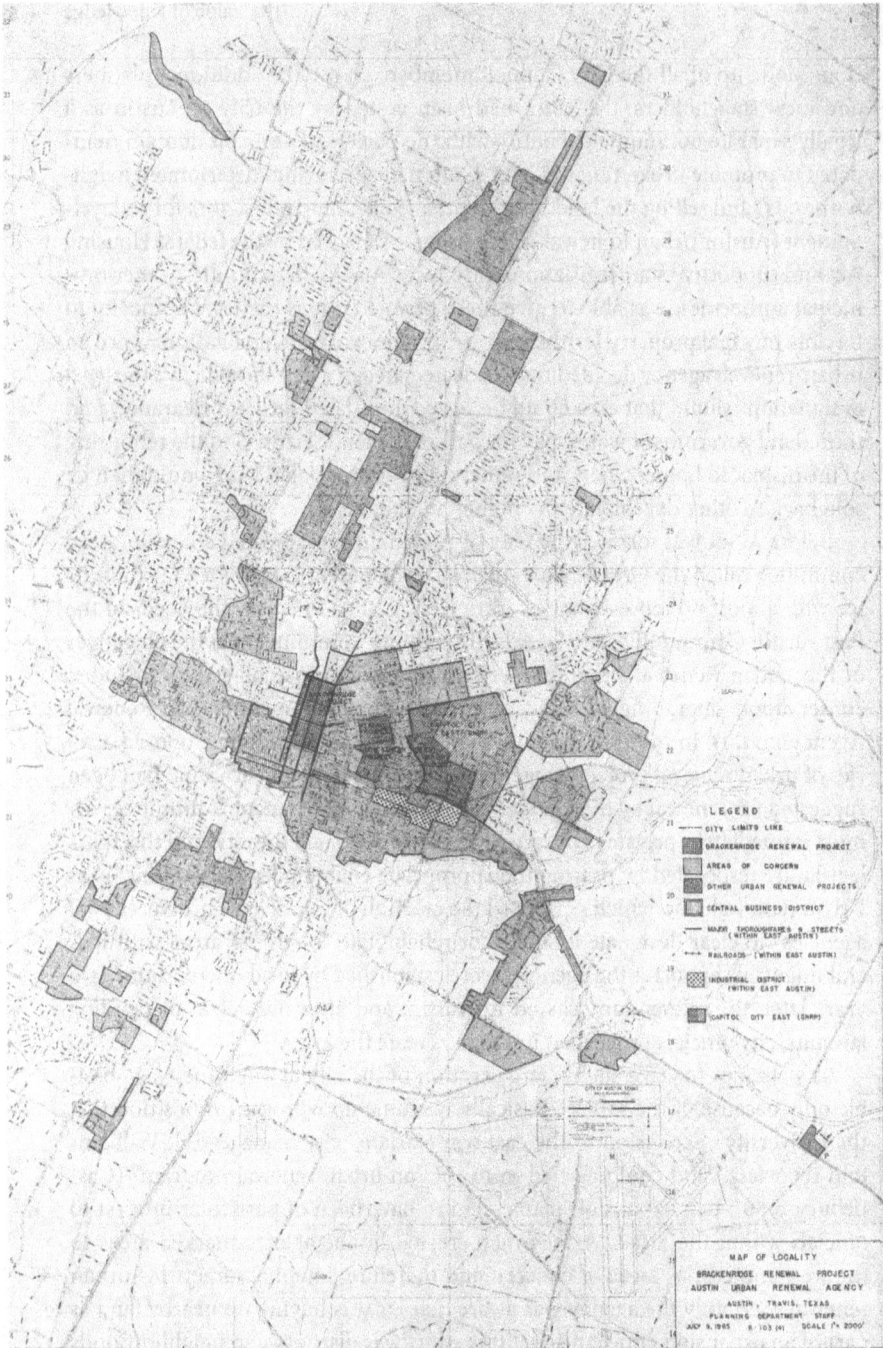

LEGEND

— CITY LIMITS LINE

BRACKENRIDGE RENEWAL PROJECT

AREAS OF CONCERN

OTHER URBAN RENEWAL PROJECTS

CENTRAL BUSINESS DISTRICT
(WITHIN EAST AUSTIN)

— MAJOR THOROUGHFARES & STREETS
(WITHIN EAST AUSTIN)

⟷ RAILROADS (WITHIN EAST AUSTIN)

INDUSTRIAL DISTRICT
(WITHIN EAST AUSTIN)

CAPITOL CITY EAST (CDRP)

MAP OF LOCALITY
BRACKENRIDGE RENEWAL PROJECT
AUSTIN URBAN RENEWAL AGENCY
AUSTIN, TRAVIS, TEXAS
PLANNING DEPARTMENT STAFF
JULY 9, 1965   B-103 (4)   SCALE 1"= 2000'

FIGURE 4. Austin Urban Renewal Agency Map of Austin with Area Designations, 1965.
The Austin History Center.

the single redeveloper for the entire portion of the land that was scheduled to be taken. Moreover, administrators at the agency would have been particularly interested in involving UT as a partner because new federal statutes in the 1959 Housing Act allowed them to take advantage of the special two-to-one credit program (which they eventually did). For the regents, the plan would help alleviate their main concern, which was the potential financial costs of such a large land seizure and the additional money required to move the facilities from the old campus or construct new facilities on any newly acquired area; if the federal government accepted the plan, it would contribute two-thirds of the net costs of taking land in the Blackland (University of Texas Board of Regents 1966a). An additional benefit to acquiring the land with AURA's assistance was that doing so helped manage the expansion's social costs by having the federal government cover the expense of rehousing displaced people. Had the regents wantonly exercised their power of eminent domain, none of these costs would have been covered, a point stressed by members of the City Council that supported the federal government role in the university's land clearance program. However, it is unclear if such a huge land seizure proposal would even have been possible without sources to assuage the pain for the affected communities (City of Austin 1966).

The full extent of the eastern proposal, however, was never realized. Almost a year after the original proposal, in February 1967, the university announced that it was going to acquire only a little more than half of the originally planned 140 acres (Castlebury 1967). The reason, according to McCarver, was that "city officials could not find enough dilapidated housing to qualify for HUD (federal) funds" (McCarver 1995, 26). Instead, the university, attempting to gain more land, joined a plan related to another proposed urban-renewal project area (known as the Brackenridge), southeast of the main campus, that mostly involved the expansion of the state government's facilities (Austin Urban Renewal Agency 1967). The regents had originally rejected participating in this project area because less land was available to them but also because there would not be any federal subsidy (University of Texas Board of Regents 1966a). However, the final plan submitted to the federal government by AURA did ask for federal subsidies to cover two-thirds of the land acquisition costs, and at least $800,000 was provided to help displaced individuals and small businesses with their moving and relocation costs (Austin Urban Renewal Agency 1967, 1972). Nevertheless, the property that UT acquired within the Brackenridge Project area was only about thirty acres, at a cost of about $4.5 million, and several of these acres could not be redeveloped because they contained historic buildings (Alcalde 1968). Although properties in other areas of the tract were slated for rehabilitation, as a concession to UT, on the land acquired by the university, all the other structures, regardless of their condition, would be razed. This condition drew intense criticism from the surrounding community (Dupont 1969). Despite these limitations, by 1975 UT still managed to acquire over one hundred acres with the assistance of AURA and the federal government.

These land clearances were devastating for the residents of the affected communities, particularly African Americans. As a 1984 exposé in the *Austin American-Statesman* noted, "For many owners urban renewal engendered bitterness that lives on, bitterness aggravated by city government inaction, lawsuits from angry neighborhood groups, and tragic accounts of families who lost the homes where they and their parents were born" (Pinkerton 1984a). In the two urban renewal areas taken over by UT, approximately one thousand people were displaced, mostly African Americans, and countless businesses, many African American–owned, were forced to close. While many people received relocation funds, the money was insufficient for people to buy a new home (Pinkerton 1984b). The result was that even more African Americans settled in East Austin because it was the only place they could afford to live, and these communities became much more isolated and increasingly disadvantaged (City of Austin 1979; Wilson, Rhodes, and Glickman 2007). Moreover, the urban renewal projects were plagued with problems. Housing projects were delayed, people went without payment, and many people found themselves moving from one area to another only to find that the new area was being condemned by urban renewal and that they would have to move again (Cryer 1969).

Conclusion

The failure to acquire a large tract of land to the east of the main campus did not stop UT from pushing east, although these later efforts would be without the support of the local urban renewal authority. In 1981 the regents asked UT president Peter Flawn to "take all necessary steps with regard to the purchase of additional acreage adjacent [to] or near" the main campus "for the purpose of future campus expansion" (University of Texas Board of Regents 1981, 125). Shortly after, the university announced it was going to take about twelve acres of land in the Blackland for the relocation of physical plant facilities from the main campus. Public ire once again emerged but eventually evaporated; the public was unaware that at the same time the university, through a third-party real estate broker, had begun secretly buying up additional plots in the Blackland that were adjacent to the 1966 and 1981 takings. After the first acquisitions happened, the administration's furtive plan was revealed, and there was immediate and loud public opposition. A nearly decade-long fight between UT and the neighborhood's residents ensued, receiving the attention of national media outlets such as the *New York Times* and the *Chronicle of Higher Education*. Unlike in the 1960s and 1970s, this time local community activists and student organizers thwarted the bulk of UT's expansion plan (McCarver 1995).

It is important to recognize how much the university's land-development efforts reflected the university's mission at that time. In the 1960s and 1970s, UT remade the area around the campus and downtown Austin because of the federal government's growing involvement in higher education, particularly for research and development. UT "needed" to improve its research profile in order to capture more federal funds. Significantly, while university professors often partnered with various industries and supported industrial development, their research and the research profile of the university were largely financed by the federal government. Therefore, the role of the university was ancillary; it simply acted as a siphon for federal funds that could have positive spillovers for regional development. The landscape built in the areas UT acquired was designed to support the development of the inner workings of the main campus, particularly helping elevate the university's research profile, and largely consisted of support services, a series of specialized libraries (such as the LBJ Presidential Library), and conference centers.

Today there is a new vision for research universities, and this vision is reflected in UT's redevelopment plans for much of the land it acquired under urban renewal in the 1960s. The latest vision for UT and other universities imagines these institutions of higher education not as just partners with industry but as industrial entrepreneurs. University knowledge products make up an ever greater share of university's income, knowledge, and money (Lawton-Smith 2006, 124–128). Moreover, there is a growing belief that universities should collaborate with industry in knowledge production, which is why the private sector provides a much greater share of university research funds than forty years ago (Atkinson and Blanpied 2008, 41). Additionally, university administrators contend that the knowledge a university produces should be, at least in part, the commercial and profitable property of that university; this trend is especially true in the biomedical sciences (Geiger 2004, 216–230). Presently, the area of Austin that was seized as part of the Brackenridge Urban Renewal Project is being remade into a medical district, and there are plans to create an innovation district in the neighboring area. The medical district is being built by a partnership between UT and two health care providers (Seton and Central); it also involves a sizeable contribution from computer mogul, Michael Dell, who briefly attended UT as an undergraduate from 1983 to 1984. Initially, the area will have a new medical school, a research building, and a new teaching hospital run by Seton. Later, an entire campus dedicated to research, particularly biomedical research and technology, is slated to be built. Moreover, there is hope that the construction of an adjacent innovation district will "provide space and technical resources that can leverage university research and discoveries into new businesses and products" (University of Texas at Austin 2013, 36).

After the landscape is completely refashioned to make way for the new medical facilities, few people will probably miss or want to remember the rather

barren landscape the university built to its southeast in the 1960s and 1970s, a
landscape whose crowning monument was a hulking, modernist concrete cyl-
inder named the Frank Erwin Special Events Center, a tribute to the chairman
of the Board of Regents who supervised the land expansion. But it is important
to remember the tragic effects of these land seizures that displaced hundreds
of underrepresented peoples. Most Austin residents remain ignorant about the
sacrifices these people were forced to make to pave the way for a "city of knowledge."

CHAPTER 3

Bringing In the State

State Entrepreneurialism, the University, Land Development, and the High-Technology Sector

The battle for national leadership among states is being fought on the campuses of the great research universities of the nation.

Peter Flawn, President of UT Austin

The presence of these major international companies has made Austin a global center versus a regional one, and the effect is magical.

Angeles Angelou, former Vice President of Economic Development of the Greater Austin Chamber of Commerce

Over the last forty years, scholars have noted how state governments have increasingly taken an entrepreneurial stance toward the promotion of industrial development (Eisinger 1988, 1995; Fosler 1992). State governments used their institutional powers and resources to propel industrial development for most of the twentieth century, but since the 1970s the strategies states have used to promote economic growth have become much broader and have taken on new characteristics (Cobb 1993). In particular, there has been a noticeable shift to what Eisinger described as supply-side vs. demand-side interventions. The older system centered on ways a state could amplify a business's effects on regional development through supply-side actions that lowered the costs of inputs through methods such as tax abatements. In contrast, more and more state governments have become willing to use demand-side policies and programs focused on developing new market opportunities through programs such as export assistance, venture capital funds, or technology transfer agencies (Leicht and Jenkins 1998). Moreover, newer state development strategies have focused almost exclusively on recruiting or promoting industrial development in newly emerging high-technology sectors of the economy (Eisinger 1988, 266–289; United States Congress 1983, 7).

These state industrial strategies also increasingly targeted the research functions of universities as resources that states could use to promote economic growth in high-technology industries (Eisinger 1988, 275; Feller 1992). Since at least the 1970s, state governments had supported certain kinds of university research in the hope that these investments would create or maintain an educated workforce that could attract or spawn new high-technology firms (Long and Feller 1972). State governments had also shown an ever greater willingness to help finance and develop research parks to promote profitable collaborative relations between businesses and public universities (Luger and Goldstein 1991, appendix A). However, by the 1980s, as states created a much broader array of policies and programs to support technological development, these efforts, such as creating agencies that helped patent or license university discoveries, also focused on new ways to use institutions of higher education to promote regional development (Peltz and Weiss 1984; Schmandt et al. 1987).

The rise of this new entrepreneurial state paralleled and was connected to a transformation in how universities, particularly administrators and faculty, approached the value of their research programs (Etzkowitz 2003; Slaughter and Rhoades 2004). Increasingly, public and private universities and their faculties were getting into the business of directly profiting from their research in varying ways, from licensing agreements or patents on new technologies to having an ownership stake in newly formed private spin-off enterprises that resulted from university-business collaborative partnerships (Geiger 2004). In public universities the pursuit of these revenues may be more connected than in private universities to the decline in state support, but this change occurred in all major research universities. Moreover, federal funding for all university research declined from about 67 percent in 1975 to about 60 percent in 1990. This decrease was largely offset by an increase in funds from private industry, whose contributions to university research rose from about 3 percent to about 7 percent over the same period (Atkinson 2008, 220). These funds, while small, are nevertheless important because they were designed to encourage closer university-industry collaborative relationships in order to ease the transfer of potentially commercializable knowledge from universities to the private sector (Berman 1990; Fairweather 1990). In many cases, these collaborations resulted in the development of new and immediately financially profitable products or process innovations and provided new revenue streams for universities (Lee 2000).

Yet this scholarship has largely missed the growing significance since the 1970s of universities as land developers. In particular, the research overlooks how these land development efforts were connected to how universities embraced changes in their research programs to support regional growth and new revenue streams (Levitt and Porter 1985, 4). Increasingly, universities began to use their endowments to buy additional land holdings, in part to diversify their financial portfolios, and to develop existing land into research parks or other commercial ventures to gain revenue from leases (Vermeulen 1980; Fink 1983). Universities were entering into the market as land speculators and, in many cases, were seeking to capitalize on their reputation as research centers in order to attract new tenants in high-technology sectors such as biotechnology or microelectronics (Levitt and Porter 1985, 21; Pinck 1993). Moreover, through land development projects, universities could support regional economic growth by developing the physical infrastructure needed to attract or support the creation of new businesses, and they could use their resources to support research, often in collaboration with the private industries located in these facilities, to create new spin-off firms (Premus 1985, 24–33). While private universities were often at the forefront of these efforts, public universities also tried to establish new ways of gathering revenue from land development projects (Levitt and Porter 1985, 12–21). However, as quasi-state institutions, state universities often have special privileges, such as the ability to issue low-interest public bonds or a general exemption from any local land-use controls or zoning regulations (Curzen and Lee 1985, 60; McGown 1985, 46).

In the following chapter, I argue that Texas's industrial strategy to promote the regional development of high technology in Austin was inextricably connected to UT's unique role as the state's agent in land development. In the late 1970s UT was at the forefront of figuring out new ways to harness the economic value of the university's research to propel regional growth (Oden 1997; Smilor et al. 2007, 210–214). In particular, it developed an effective internal institutional system that eased technological transfer and encouraged profitable commercialization of discoveries; these efforts were firmly joined to a host of new state government initiatives in the 1980s (Butler 2010). Moreover, although UT had been a strong part of the local urban growth coalition for most of the city's history, during the 1980s the university's role within the local economy was enhanced, as was its power, in part because of the state's efforts. The state government helped refashion UT, as its agent, into a land developer that could support the industrialization of central Texas. To be sure, other state governments had been active in engaging their institutions of higher learning as the state's agent to support regional growth through land development, most significantly in North Carolina's research park. However, UT's actions stand out because it, as part of the state's program to promote industrial development, used its unique bonding authority, discussed in chapter 2, solely to acquire and develop land for two nationally recognized high-technology firms. Its role as land developer, in fact, proved to be the critical incentive in the state's successful attempt to secure this investment, an investment that drove Austin's regional development for decades.

The following chapter is broken into three sections. The first reviews the literature on the role federal and state policies have played in supporting the commercialization of university research across the United States and the success of other universities, particularly Stanford, in promoting growth in their regions, in part because of their role as land developers. This is followed by a section that examines how Austin began as a center for high technology, particularly the central role UT played as a land developer in securing the investment of two national corporate high-technology consortiums, and how having these firms repositioned the city within the national urban hierarchy. Finally, the last section shows how a new configuration of actors in the local growth coalition grew out of these efforts and how the state of Texas, because of UT's enhanced position as its actor, became a more significant force in driving economic growth in the region.

Federal and State Policies and the University as a "New Breed" of Land Developer

Beginning in the 1980s, federal policy toward science and technology increasingly became organized around a new "competitiveness agenda." The centerpiece of the agenda was to use "government funds to commercialize science and technology" in order to "increase U.S. share of global markets and to increase the numbers of high-technology, high-salaried jobs in the domestic economy" (Slaughter and Rhoades 1996, 304). The new policy grew out of the efforts of a coalition of interest groups that emerged in the late 1970s, led primarily by the emerging "medical-industrial" sector of biotechnology and pharmaceuticals. Within this sector, agents driven to increase their competitive position in the international marketplace worked with university managers (and some faculty, especially in the material and natural sciences) to steer federal research policy toward commercializing all research (including research done at universities) and establishing a system of government subsidies for that research. This newly formed coalition was instrumental in persuading national legislators to adopt new federal laws, such as the Bayh-Dole Act of 1980 (often likened to a second Morrill Act, an act discussed in detail later), that veered away from long-standing precedent and allowed all entities that received federal funds for research to patent, commercialize, and profit from their discoveries (Henderson and Smith 2002; Slaughter and Leslie 1997, 316–319). Slaughter and Rhoades argue that the changes brought about under the new competitiveness agenda also affected how universities were managed and governed. In particular, they argue that this new system of "academic capitalism," which orients university administrators increasingly toward a profit-centered outlook, emerged because of the desire to exploit the potential economic returns that could flow from more seamlessly embedding universities into the productive structure of the new knowledge economy.

The work of Slaughter and Rhoades is just one in a myriad of books and articles on the rise of what could broadly be called the "entrepreneurialization of the university" (Brint 2005; Etzkowitz 2003; Etzkowitz et al. 2000; Newfield 2003). All of these accounts, while different, seek to identify, and often to criticize, the reorientation of the university's core functions toward seeking profitable avenues for research. While these changes have had various effects on university education, research, and management and have been experienced differently across a range of disciplines and specializations, one noticeable shift is the increasing willingness (and need) of universities to open new revenue streams from research discoveries either by partnering with the private sector in business ventures or through licensing and patenting agreements (Geiger 2004; Berman 2011). While these partnerships have primarily occurred in engineering and the natural and health sciences, business and management schools have also been increasingly important in building more financially profitable areas of research such as producer or business services (Slaughter and Rhoades 1996, 308).

The ability of university managers to partner with the private sector has largely been made possible by changes in the potential commercializability of basic research for the university. As Slaughter points out, "Change in the economy and the structure of science offered the university an alternative source of funding. In the early 1970s, the corporate community became enamored of high technology as a way to improve its position in the global market. At the same time, university science became a source of high technology. Discoveries in biotechnology, pharmaceuticals, computers, lasers, material sciences, and robotics made basic research performed in the university more available for product development and marketing. The distinction between research and development, always somewhat artificial, seemed to be disappearing entirely" (Slaughter 1990, 43).

The distinction between research and development (or between basic and applied research) had by and large reflected demands the federal government and private industry had placed on universities in the 1940s (Newfield 2003, 122–123). "Basic" (or fundamental) research came to be defined, particularly in material and natural sciences, as research that had no immediate commercial application. With the massive expansion of universities in the second half of the twentieth century, more of this type of research was undertaken largely by university researchers. "Applied" research had immediate uses and was done in research labs run by private companies or facilities, often in research parks, managed in partnership with universities.

Of course, this did not mean that "basic" research conducted in universities before the development of the knowledge economy was not valuable to government and commercial and private interests. Basic research always had a recognizable value for national defense and armaments, and this value in part drove universities and industrial firms into more intricate collaborations immediately after World War II (Kargon, Leslie, and Schoenberger 1992; Saxenian 1996). Moreover, private firms were interested in universities' research practices at least as far back as the early twentieth century, in part because of electrical and chemical firms' success in making innovative and profitable discoveries by funding research in private facilities (Noble 1977, 110–128). In fact, industry has been engaged in collaborative relationships with universities since the end of the nineteenth century, and this practice became increasingly common at the beginning of the twentieth century (Geiger 1986, 174–177; United States National Science Board 1983, 1–4). At the dawn of the twentieth century, private industry was increasingly willing to finance research in universities because doing so served at least three purposes: (1) universities could produce more well-trained technical experts; (2) private industry could exercise a large degree of control over what was researched in universities; and (3) this arrangement "successfully shifted the burden of some significant costs, and risks, of modern industry from the private to the public sector" (Noble 1977, 147). The last point is especially important because many of the nation's leading research universities, especially those that expanded dramatically in the 1950s and 1960s, were public universities that continued to receive large portions of funding for their physical plants from state governments, even though private industry provided the majority of funding for research. In the period after the Second World War, however, private support for basic research became less significant to universities as the federal government's financial support of basic research grew dramatically (Graham and Diamond 1997, 34).

Many of the largest U.S. research universities are public and, therefore, creatures of their states, but federal policy has played a unique role in their development. Beginning in the early nineteenth century, the federal support to universities grew dramatically because of its role as a grantor of land. In the latter half of the nineteenth century, a series of federal legislative acts (perhaps the most famous being the 1862 Morrill Act) created a legal framework to provide land grants or a scrip to states for the development or expansion of institutions of higher learning in the "agricultural and mechanical arts" (Brubacher and Rudy 1997, 154). Many of today's leading public, and a handful of private, universities were established as a result of federal land grants. Later, in the second half of the twentieth century, the federal government greatly expanded its role in higher education by providing direct financial support for undergraduate and graduate education as well as faculty research (Clark 1991). Moreover, beginning in the 1970s, the federal government, largely through the National Science Foundation, took a leading role in helping foster closer relationships between universities and the private sector (Bowie 1994, 15–22).

Despite significant federal involvement, U.S. public higher education systems are the providence of state governments. For decades, state governments have been the primary vehicles for financial support, particularly for undergraduate education at state universities, as well as providing oversight in the management of public universities (Eckel and King 2007). Moreover, state universities were, in part, established with the intention of a more productive and cooperative relationship between the private sector (agriculture and industry) and the university (in both teaching and research) (Adams 2002, 275). In many instances, these collaborations were intended to enhance and cultivate particular local or regional geographical advantages, either industrial or agricultural (Cohen, Florida, and Randazzese 1996, 52–53). Since the 1970s, as more and more state governments created programs and policies to support the development of technological innovations, states across the country began to engage universities as indispensable partners in state economic growth strategies (Schmandt and Wilson 1988; Lambright and Tefch 1989; Harloe and Perry 2004; Lendel 2010; Mayer 2007; Luger 1985, 213). Programs were developed to garner outside investment or support indigenous firms, especially ones involved in research-intensive developing technologies, by exploiting the special relationship between states and their public universities (Feller 1988). The hallmark of these approaches was to harness the potential economic value of university research by promoting closer collaborations with the private sector in order to produce viable spin-off firms, or to encourage the direct commercialization of university discoveries through patenting and licensing efforts (Peltz and Weiss 1984).

To a large extent, state strategies developed in the 1980s that focused on university research and its potential returns for regional growth sought to emulate the economic success of certain regions, namely Silicon Valley in California, the Route 128 corridor in Massachusetts, and Research Triangle in North Carolina (Washburn 2008, 175). In these locales, the close and collaborative relationship between universities (particularly Stanford; the Massachusetts Institute of Technology and Harvard; and Duke University, the University of North Carolina, and North Carolina State) and private businesses had helped propel the areas' economic development (Smilor et al. 2007; Graham and Diamond 1997, 120; Fogarty and Sinha 1999). In particular, Stanford and MIT stood out because they had played a strong and active role in their regions since the 1950s, often acting like entrepreneurs by supporting the development of specific private firms.

The importance of universities to the development of industry in Palo Alto and Boston had been highlighted by public officials in the 1960s, and in the second half of the twentieth century there was an increasing recognition that universities had a key role in the development of what Margaret O'Mara has aptly called "cities of knowledge" (O'Mara 2005, 90–92). According to O'Mara, these cities, which had become the new centers for industrial innovations, shared at least three characteristics: (1) newly built and extensive suburbs were filled with predominately white, wealthy, and educated people; (2) a large number of these people, mostly white men, worked in firms in a research or management capacity; and (3) many of the so-called growth nodes in these suburbs were connected to technology firms that often had been developed with the financial support of universities. Beginning in the 1940s, Cold War ideas about planning stressed the necessity of "deconcentrating" vital infrastructural resources, particularly those associated with national defense, because of the danger posed by a focused missile attack on overconcentrated resources. Since most high-technology firms were part of the national defense infrastructure, their suburbanization was encouraged as part of a dispersion policy that dictated how defense-related funds were spent (O'Mara 2005, 18–57). Moreover, local urban elites, aware of the opportunities this spending offered for urban growth, pushed to capture a portion of the federal largesse, especially for infrastructure. So too did university administrators, who believed that the growth potential of their universities was increasingly connected to their ability to capture large portions of federal spending on research, especially in areas related to defense.

It is no surprise, therefore, that universities began to become "a new breed of city builder" (O'Mara 2005, 59). In particular, as universities became more engaged in urban and regional development, especially as land developers, their efforts helped foster the growth of many suburbs. The embodiment of their newfound role as land developer was the myriad research parks built in the 1960s, often in newly developing suburban communities. Largely modeled on the suburban industrial parks of the 1950s, research parks were unique "in their relationship with institutions of higher education" (O'Mara 2005, 69). Usually managed and owned by universities, research parks offered a space to connect the priorities of private-sector industrial development to the research capacity of universities. While they often housed private-sector laboratories, many research parks were indistinguishable from the university's main campus. Moreover, firms in research parks often enjoyed a close relationship with university departments and faculty, in some cases even employing faculty to work as consultants.

Since the 1970s, more and more cities and regions have attempted to become "cities of knowledge," and some, like Austin, have been successful. In part this reflects the increasing significance of the commercial value of university research for regional economic growth (Harloe and Perry 2004; Lendel 2010; Mayer 2007). It also shows how many more universities became more active in developing land and providing the critical physical infrastructure improvements to support the development of high technology in their regions (Luger and Goldstein 1991). In fact, the two activities increasingly converged. Universities increasingly sought out new ways to tap the financial returns of their research capacities, often conjoining these efforts to an ever greater number of real estate ventures that sought to retain, promote, or attract high-technology development (Premus 1985).

However, universities were not alone in supporting efforts to remake their function as producers of knowledge into a core driver of economic development. As the section has shown, the federal government and state governments enacted essential legal and policy changes and established programs that connected the research functions of universities to industrial development policy. These policies and programs were significant in helping universities capitalize on and reinforce the broader changes in the economy that increasingly valued knowledge-intensive practices. As I will show, state policy was also critical in refashioning the role that a university would have as a land developer and as the state's agent in promoting urban growth.

The Making of a High-Tech Austin

Austin's pattern of development is a paradigmatic case of a "city of knowledge." High-technology firms have tended to locate in the city's growing northwest and southeastern suburbs since at least the 1960s, and this process, which accelerated during the 1970s and 1980s, was largely reinforced by the expansion of UT's suburban facilities (Kahn and Farley 1984, 91; Ryan, Horton, and Arbingast 1968, 1; Brooke 1980).

Seeking to take advantage of the external economies of scale, high-technology firms were located in distinct clusters along U.S. 183 (Research Boulevard), a road built in the 1950s that runs from the northwest to the southeast section of Austin. One area of considerable development was in the northwest section, at the intersection of Highway 1 (the MOPAC Expressway) and U.S. 183, where UT's Balcones Research Center (discussed in the next section) was located. By the early 1980s a sizeable agglomeration of technology firms, including IBM, Radian, Texas Instruments, and the Microelectronics and Computer Corporation, could be found in this northwestern quadrant. Southeast along U.S. 183, there was a cluster of firms near the intersection of Interstate 35 and another large one further east at Farm-to-Market Road 969. There, defense contractor Tracor built a large manufacturing plant in the 1960s, and Motorola and Rolm established facilities in the 1970s and 1980s, respectively. Finally, at the intersection of U.S. 183 and State Highway 71, where Sematech would later be located, there was a considerable group of technology firms that included, among others, Advanced Micro Devices, Lockheed, and Data General (a company that later sold its land and buildings to UT).

As in the rest of Texas, the development of the high-technology economy in Austin is connected to the largesse of federal military spending (Farley and Glickman 1986, 409). One local Austin firm that benefitted from this money was Texas Research Associates (Tracor), a defense electronics contractor founded in 1955 (Kleiner 2013). The first Austin-based company listed on the Fortune 500, Tracor played an especially important role in the region's development because by 1975, the company had spun off at least fourteen other high-technology firms (many working on non-defense-related products) in the region.

Moreover, the company's success might also prove the value of having a close relationship with the University of Texas. Tracor was formed as a partnership among four researchers at UT's Defense Research Lab (DRL): mechanical engineer Frank McBee and physicists Richard Lane, Chester McKinney, and Jess Stanbrough. The first collaboration of its kind in Austin to effectively commercialize on this research relationship, "TRACOR drew its talent almost exclusively from University of Texas engineering departments and especially from the DRL" (Busch 2011, 166). Throughout its history, Tracor and its spin-offs (and their spin-offs) would use both informal and formal connections to the university to boost their fortunes, particularly by tapping into the knowledge and scientific expertise of UT faculty and students. One notable instance was chemistry professor and former UT president Norman Hackerman, who joined the company in 1956 (Busch 2013).

Tracor was not the first firm in Austin to benefit from a connection to the university. Beginning in the mid-1940s, a growth coalition led by the local automobile dealer C. B. Smith crystallized, mainly through the work of the Industrial Division of Austin's Chamber of Commerce, which had been established to figure out a local strategy to increase Austin's industrial infrastructure (Austin Area Economic Development Foundation 1948). The members of this group included local businessmen such as bankers and realtors, leaders from local government, and one chemical engineering professor from UT, William Cunningham. Government and educational services, whose economic returns were mostly seasonal and also lacked many of the compounding returns found in industrial sectors, still overwhelmingly dominated Austin's economy. Part of the development strategy offered by the coalition relied on maximizing the central role UT could play in supporting private industrial research. As an article published in 1950 noted, "Their plan particularly encompassed the encouragement of new enterprises which could use the University's research facilities for the development of their products and made a contract with the University for that purpose, probably the first such arrangement ever entered into between an institution of learning and a civic organization" (C. B. Smith 1984).

The group then proposed creating the Austin Area Economic Development Foundation (AAEDF), as a subsidiary of the Chamber of Commerce, to coordinate efforts between UT and businesses in the region. In addition, the AAEDF also helped connect upcoming industries or firms that decided to locate a branch in Austin to loans and other financial resources.

Additionally, the UT-controlled Balcones Research Center (BRC, now renamed the J. J. Pickle Research Campus) played an important role as an interface between the business community and the university. The facility had been a magnesium factory during the Second World War, after which it had been decommissioned and remained unused. Two engineering professors, C. R. Granberry and J. Neils Thompson, strongly supported by Austin's postwar growth coalition, spearheaded the effort to acquire the property for research purposes, and after a series of lengthy negotiations with the federal government, many involving then congressman Lyndon Baines Johnson, UT was able to secure a long-term lease for the site (Long 1962). The plant was then quickly refashioned as a research site, primarily for defense-related engineering research. While the BRC had a minor role in the 1950s in helping Austin secure the investment of some private firms, such as Jefferson Chemical, the site remained exclusively dedicated to noncommercial university research, particularly research that was associated with national defense.

Notable changes happened at the BRC in the 1970s, however. First, UT gained title to the land in 1971, when the federal government gave the property to the university at no cost. After this transition, at the behest of UT's regents, the BRC was reorganized "as a campus for substantial university research facilities in interaction with industry and government" (Kragie 1986, 21). To support this new mission the BRC underwent several substantial land expansions and became a center that "pioneer[ed] new ways for The University and industry to work together . . . in areas of advanced research and development" (Kragie 1986, 22). However, unlike some other research parks (most notably Stanford's), the BRC did not became a site for the operation of a large number of for-profit companies. Nevertheless, the BRC was successful in supporting local industrial research through shared collaborations, and this, in part, helped propel Thompson's lifelong interest in figuring out ways to forge a more profitable relationship between the university and the business community. Thompson later served in several leadership roles at Austin's Chamber of Commerce, including as its president from 1973 to 1974 (Busch 2011).

Austin had developed a small but significant infrastructure of high-technology firms by the early 1980s, but their impact on Austin's urban development remained relatively limited. At the start of the 1980s, most of the firms were homegrown, and research-related firms were directly associated with or spawned from Tracor or the branch plant locations of several national companies, including Texas Instruments, IBM, Motorola, and Rolm (Brooke 1980; Hobbs and Rundell 1982; Gibson, Smilor, and Kozmetsky 1991; Glasmeier 1988).

However, by the end of the decade, the number of technology firms in the city had grown at a phenomenal rate, in part because of decisions by two major national research firms, Microelectronic and Computer Corporation (MCC) and Semiconductor Manufacturing Technology (Sematech), to make Austin the location of their national headquarters. Numerous studies, particularly the exhaustive book by Gibson and Rogers, have investigated how these firms helped change the composition of the regional technology infrastructure and provided a catalyst for the city's explosive growth in the 1980s. There is no point in repeating all of the details, so instead the next section focuses mostly on the two aspects that are major themes of this chapter: the role of the state government and UT's role as a land developer.

There were at least four important similarities shared by MCC and Sematech, which are explored in more detail later in this chapter. First, both were established to help increase U.S. competitiveness in R & D, particularly in the electronics industry, at a time when the nation's position within the international hierarchy was, many argued, connected to innovation in core industrial sectors (Kahn and Farley 1984, 1–2; United States Congress 1992, 2–3). Second, both were examples of a new type of private for-profit company, a for-profit research consortium, where firms in similar, if not competing, sectors of the economy were able to legally collaborate on basic research projects. Third, to select the

site for their national headquarters, both corporations held nationwide compe-
titions, which enticed many locales and their states to make very generous offers
of subsidies and incentives to attract their investment. Finally, and perhaps most
significantly for my argument, UT played a critical role in helping Texas secure
both firms through its ability to use its unique bonding authority to become a
land developer.

MCC was the first for-profit research consortium of its kind and was established to help rebuild the competitive position of the United States in the microelectronics sector by undertaking basic research that had no immediate commercial applications (Frontain 2010). In 1982 the antitrust division of the U.S.
Justice Department, after an intense lobbying effort by MCC and its supporters,
deemed such industrial collaborations through consortiums legal, in part because it held that existing antimonopoly provisions were unnecessary restrictions that only fettered U.S. industry in the face of increased international competition. In 1984 the principle behind the Justice Department's decision became
law after Congress unanimously passed the National Cooperative Research
Act (NCRA) (Gibson and Rogers 1994, Rule 1985). Several large private share-
holding companies, which owned varying stakes in the firm, financed MCC's
multimillion-dollar annual operations, but these shareholders did not receive
direct monetary dividends. Instead, MCC paid back its investors with basic research discoveries that could be used by any of the collaborating firms for commercially viable ends or to help create separate, profitable spin-off companies.

In 1987 Sematech, an industrial consortium of semiconductor firms, was
formed (Browning and Shetler 2000). Like MCC, Sematech brought together
leading national firms in similar areas of the economy to collaborate on basic
research. However, unlike MCC, the federal government had an active role in the
consortium as a financial partner. In fact, half of the firm's $200 billion annual
budget came from its fourteen shareholder companies, and the other half came
from the federal government's Defense Advanced Research Projects Agency
(DARPA). The government was so heavily involved because, many influential
figures argued, declining U.S. competitiveness in semiconductor technology
posed not only an economic problem but also a substantial defense risk (Gibson and Rogers 1994, 475–478). So while Sematech, like MCC, engaged in basic
research that had no immediate commercial applications, its primary goal was,
as Browning and Shetler noted, "to improve core manufacturing competence
rather than on making breakthrough end products with competitive features."
Still, over time, some discoveries made at Sematech did become commercially
viable products or processes, and several nondefense firms were profitably spun
off (Browning and Shetler 2000, viii).

UT, particularly the bonding authority provided by the Permanent University
Fund (PUF, discussed in chapter 2), had a crucial role in getting MCC to locate its
national headquarters in Austin. In 1983, fifty-seven cities in twenty-seven states
competed to be the location of MCC's corporate headquarters; Texas, in conjunc-

tion with Austin, was selected primarily because it was able to offer the largest incentives package of any of the competitors—a package worth more than $70 million. Importantly, about $50 million came directly or indirectly from the UT system. UT justified such a large expenditure because of the potential benefits that MCC would provide the university. By 1989 it was estimated that MCC contributed about $1.7 million annually to the university, and the largest chunk of these funds, about $1 million, was provided as grants to UT researchers (University of Texas Board of Regents 1988e; Benningfield 1989).

Originally, about half of the money for MCC was supposed to come from UT's own funds or special Texas legislative appropriations to the university, and the other half was supposed to come to the university from private sources that made donations to "The University of Texas System High-Tech Fund" (Downing 1983). The largest single portion of the state's subsidy for MCC, around $20 million, was for the construction of a new building on an undeveloped tract at the BRC that was worth an estimated $3 million but was to be leased back to MCC for a mere two dollars a year (University of Texas Board of Regents 1983). At first, UT's regents had agreed to pay only $5 million (it later ended up being over $14 million) of the $20 million for the new building from any of its sources, including proceeds from PUF bonds that could legally be used only for capital improvements (University of Texas Board of Regents 1984, 148–149). Approximately another $27 million, either from the dividends derived from or the sale of bonds issued against the annual revenue of the PUF, depending on the expenditure, was offered: $15 million (of a $20 million pledge from the state government) to help the departments of engineering and computer science recruit new, or give raises to existing, faculty; $2 million to purchase new equipment for MCC; and about another $10 million (given as a special legislative appropriation) for various projects, such as supporting the research of graduate students in computer science and engineering. The remaining $23.5 million in subsidies promised to MCC was supposed to be raised from private donors by the Texas High-Tech Fund: $15 million, later increased to over $16 million, for the costs of construction and about $7 million for faculty support. While the fund was established at the behest of the governor, it was spearheaded by a group of Texas's business elites, who helped raise private contributions and funnel them through the university to MCC. Its fundraising goal, however, fell short by about $7.6 million, and eventually UT was forced to pay an additional $9.5 million from its bond proceeds toward the construction of MCC's new office building (Benningfield 1988; Gibson and Rogers 1994, 155; University of Texas Board of Regents 1987, 7). Other than the university's own funds and the private contributions that flowed through it, the only other major contributor was the local business community, primarily banks, which offered more than $20 million in additional incentives to future MCC employees; the bulk of this money was for subsidizing single-family mortgages at two percentage points below the market rate.

Sematech, like MCC, had no official connection to UT, but once again UT and

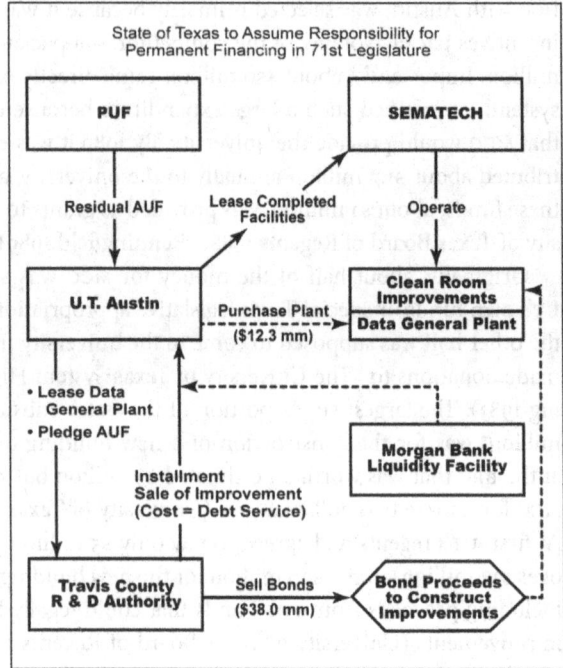

FIGURE 5.
Sematech Interim
Financing Structure.
Based on image found
in Board of Regents
Real Estate Records,
Dolph Briscoe Center
for American History
the University of Texas
at Austin.

its bonding authority had a central role as the state's vehicle to channel funds for land development. In the much more extensive national competition among thirty-four states with 135 proposals to secure Sematech's headquarters, Texas offered a similarly massive $68 million (Gibson and Rogers 1994, 487). About $18 million was to be spent in four areas: approximately $10 million on mortgage incentives, $5 million on a cooperative state research fund, $1.5 million of time on UT's supercomputer, and about $1.5 million in various other incentives. But the vast majority of the incentive package was $50 million for land development and acquisition, which, save a relatively small contribution of $250,000 from the City of Austin, came exclusively from bonds that UT issued against the PUF (Benningfield 1988; SEMATECH Foundation of Texas 1987).

The university was compelled to buy land for Sematech, retrofit an existing building, and then lease it back to the firm for a nominal fee. While UT had again wanted to use undeveloped land on the BRC site, Sematech preferred a second option that had been offered to them: the refashioning of an abandoned Data General Corporation manufacturing plant in southeast Austin. The problem was that this building was not yet UT's property, and it was unclear how the university could acquire land for a secondary user. To make it possible for UT to be the sole purchaser of land for a separate company, the Texas legislature passed a law that allowed for nonprofit research development agencies to be established between county governments and private entities, such as the university (University of Texas Board of Regents 1988b, 1988c). Subsequently, the Travis County Research Development Authority (TCRD) was established, and UT was able to acquire the property at about $12 million as an agent in this authority and re-name it the Montopolis Research Center (MRC) (University of Texas Board of Regents 1988f). The TCRD issued $38 million in bonds (the interest was paid by the university's endowment and the collateral was the PUF) to renovate the site and build and furnish a "cleanroom" (University of Texas Board of Regents 1988e).

It is worth noting that Texas's package to bring Sematech to Austin was small when compared to the offers of other states, so it should be stressed that UT's endowment played the critical role in the state's bid. In the 1986 competition many state incentive offers were as much as seven times larger than Texas's offer, but many of these proposals relied on unsecure funds such as future state appro-priations. Texas's ability to leverage UT's resources was essential, as Sematech's executives saw the funds promised in Texas's incentives package as more secure than the financial packages offered by other states (Whitney 1988; Cunningham and Jones 2013).

UT's role in supporting MCC and Sematech did not end with land develop-ment. In the 1990s, the university helped facilitate the spin-off of two for-profit firms from MCC and one from Sematech (Gibson and Rogers 1994, 452–454). These resembled earlier attempts to engage UT with the business community through the AAEDF, but in the 1980s UT had taken over the leading role in pro-moting these commercialization efforts, primarily, although not exclusively, through the Institute for Constructive Capitalism (IC²), established in 1977, and its program the Austin Technology Incubator (ATI), founded in 1989. These two institutions within UT, largely building on previous initiatives by the Chamber of Commerce, have played an increasingly important role in helping develop the regional economy by engaging UT, more and more, in activities that promote the marketability of research discoveries (Gibson and Butler 2013, 66–72).

However, the most important lesson learned from Texas's ability to secure the investment of MCC, a lesson later seen in the state's successful bid to capture Sematech, was that UT, as a quasi-state actor, could play a unique role as a land developer in supporting the local growth coalition and Texas's larger economic development strategy. As former chancellor of the UT System Dr. Hans Mark re-marked on Sematech's decision to locate in Austin: "Look at Silicon Valley. The presence of Stanford is critical (to the development of microelectronics there). The dean of the School of Engineering started it by conceiving the ingenious notion of making very favorable land deals for bright graduate students. One example was for two students named Hewlett and Packard: the board of trustees gave them a place to build a factory. It worked like crazy. We're doing the same thing. It means literally hundreds of millions of dollars for this town, but more important, it provides intellectual stimulation" (Benningfield 1989).

Of course, graduate students from UT founded neither MCC nor Sematech. Moreover, while Stanford had played a critical role as a land developer in supporting the development of high-technology firms in Silicon Valley, it leased an existing building in the Stanford Research Park to Hewlett-Packard and other local start-up firms (O'Mara 2005, 118–122). In contrast, UT built new research facilities and in one case also acquired new land for these firms. By drawing a parallel between the two cases, Chancellor Mark actually downplays the novelty of UT's behavior as a real estate developer and financier that used millions of dollars in research funds, its bonding authority, and other incentives to help secure the exogenous investment of national firms as part of the state of Texas's industrial strategy.

A New Growth Coalition:
The State, the Austin Chamber of Commerce, and UT

The decisions of MCC and Sematech to locate in Austin substantially refashioned the economic geography of the city and assisted in the broader transformation of Austin from largely a branch-plant economy into a manufacturing and commercial research and development center (Farley and Glickman 1986, 408; Oden, Kang, and Kwon 2007, 51). Although the importance of these two firms to the local economy may be overstated at times, it's undeniable that during the 1980s several high technology firms, such as AMD, IBM, Motorola, and Dell were created in, expanded, or relocated their operations to the Austin region.

Much of this growth was strongly assisted by the state government's investment in Austin, primarily through UT. For sure, UT had played an important part in the city's growth since its founding in 1881, when the decision that the university's main campus would be built in Austin delighted the business community because of the university's potentially positive effects on the city's real estate (Long 1964). The realization of the university's possible economic contribution grew more substantial from at least the early 1950s, when building an economy around science and research became a centerpiece of the local growth coalition's strategy. The state's gradually increasing investment in the engineering and natural sciences at UT supported this strategy by helping create a stable and skilled workforce that could be employed locally in firms in related fields, and it also elevated the university's reputation and helped attract outside high-technology firms to Austin (Oden, Kang, and Kwon 2007, 50; Smilor, Gibson, and Kozmetsky 1989, 53). Moreover, UT spent millions on its computer sciences and engineering programs in the 1980s, from hiring thirty-two endowed professors at salaries plus other incentives totaling $1 million each to providing

$20 million for a new microelectronics and engineering building that was developed on the west tract of the BRC near MCC (SEMATECH Foundation of Texas 1987, 7). These investments greatly enhanced the reputation and standing of the university and helped encourage further industrial growth (Gibson and Rogers 1994, 446–447).

However, what I want to suggest is that the state's refashioning of UT as its agent in support of Texas's industrial strategy fundamentally altered the university's position within the local growth coalition. The work the local growth coalition did in providing the winning bid for MCC in the early 1980s engaged the university in a new way, particularly as a land developer that could support the development of commercial enterprises. Furthermore, while the arrangement among the actors was certainly similar to those in the past, there are some noticeable and important differences. In particular, the growth coalition that supported the development of high technology in Austin was different from the previous growth coalition in at least three respects: (1) the institutional location of its key actors, (2) the relationship between the state government and the local growth coalition, and (3) the position of UT in the state's and the local growth coalition's industrial strategies.

A new political arrangement that coalesced in the Texas campaign to secure MCC can be seen in the two task forces that then Democratic governor Mark White established to help the state win the competition for the consortium: the Governor's Task Force and the Austin Task Force (Gibson and Rogers 1994, 128–134). The Governor's Task Force included elites from the local business community, the city government, and two of Texas's flagship university systems (A&M and UT). Former San Antonio mayor Henry Cisneros, chancellor of the UT System Donald Walker, and technology magnate Ross Perot, among several others, were participants. Moreover, the Austin Task Force included a mix of local elites from the private and public sectors. Six people served on both task forces, and three of these played a very significant role: the president of Austin's Chamber of Commerce, Ben Head (also president of a local bank); Austin developer John Watson; and lawyer Pike Powers, then executive assistant to Democratic governor Mark White and later president of Austin's Chamber. Moreover, George Kozmetsky, UT's dean of the College of Business, professor, and director of the IC² institute, while not officially on these task forces, played a very important role in facilitating a new relationship between UT and the business community; this was, in fact, his central mission during his term at UT. Kozmetsky, enthused at the prospect of creating a high-technology infrastructure to employ former students, helped Walker develop the "basic strategy underlying Texas' bid for MCC: raise funds to endow professorships in computer and electrical engineering . . . in order to improve microelectronics research" (Gibson and Rogers 1994, 117). Although then UT president Peter Flawn was reportedly reluctant "about involving the University of Texas in such a for-profit enterprise," Kozmetsky persuaded him and others to make the university the "cornerstone" of the incentive package offered to MCC (Gibson and Rogers 1994, 117).

When the Sematech proposal came up, this time under the leadership of Republican governor Bill Clements, the available evidence suggests that, as Pike Powers noted, "the same collaborative team was in place" (Powers 2004, 57). This time, however, the group was organized through the Sematech of Texas Foundation, an organization created by the Austin Chamber of Commerce to coordinate with the leadership of UT and the governor's office (SEMATECH Foundation of Texas 1987, 12). Again, Pike Powers, who chaired the group, played a central role in developing the incentives package to bring Sematech to Texas. Other leading members included Peter Mills, then president for economic development at the Chamber of Commerce; Edward Vetter, chairman of the Texas Department of Commerce; Jack Blanton, chairman of the UT Board of Regents; and Ben Streetman, a UT professor of electrical and computer engineering who had also played a minor role in the MCC campaign (University of Texas Board of Regents 1988a, 1988d; Barta 1996, 380). Moreover, because half of Sematech's operating budget was to come from the federal government, the power of the Texas delegation in the U.S. Senate and House of Representatives was important, especially the active lobbying efforts by then congressman James Pickle, who represented a district that included Austin (Gibson and Rogers 1994, 492–497).

There are some significant differences in the makeup of the growth coalition that formed in Austin in the 1980s and the one that existed in the immediate postwar period through the AAEDF group. First, in the effort to recruit MCC and Sematech not a single member from Austin's local government was a leading partner. Although the City of Austin did provide a comparatively small contribution to the Sematech package, nobody in Austin outside the business community appears to have been active in the recruitment effort for either firm. Moreover, only one person from the core group, which represented UT in these efforts, was from engineering or the computer sciences, even though these departments, supposed to be the primary beneficiaries of the private investment, were the reason offered for the university's involvement in the process. Instead, the people representing both universities were administrators, and the leading figure in the campaign to bring MCC to Austin was the dean of the business school. In the MCC effort, two engineering professors were consulted but had little more than a supporting role, although one of them did appear to be more active in the effort to persuade Sematech to make Austin its national quarters.

Moreover, as Austin's economic growth became connected to the state government's competitive strategy, the university was engaged in a new way, fundamentally altering UT's relationship to the state government. Farley and Glickman noted how MCC was "seen as part of a broader effort by business and government . . . MCC must be considered a significant development for the Texas economic development strategy, since it represent[ed] an unprecedented response to a global challenge through a joint venture of one of the nation's leading research universities, many of the nation's largest multinational corporations, and the state's leading business and political figures" (Farley and Glickman 1986, 409).

Sematech too was to fit tidily into the state's overall industrial policy framework. Certainly it was a major victory for the state's progrowth forces, but, more importantly, it affirmed a basic principle held by these state government leaders—UT should play a central role in boosting economic growth and supporting the diversification of the state's economy. As former Texas lieutenant governor Bill Hobby said in his official statement about Sematech's decision to locate in Austin, "There has never been a better example in recent times of the impact of higher education on the state's economic well-being" (Engelking 1996, 36).

The effect of this new state influence in Austin, particularly in the capacity of the business community to connect the state's economic strategy to local development, made it possible to largely bypass some resistance from local power structures, especially Austin's municipal government. As Farley and Glickman observed, "The development strategy [for Austin] was principally the work of the Texas business community, University of Texas administrators and state officials" (Farley and Glickman 1986, 412). Yet Farley and Glickman largely overlook the critical role played by members of the local business community, particularly those from the Austin Chamber of Commerce, which continued to have a strong role in influencing growth dynamics in Austin. These local actors were especially successful at promoting regional growth because of their ability to mobilize the new possibilities created by state intervention (particularly those associated with UT's role as the state's agent and its financing and land development capacities) and coordinate them with local business interests.

Conclusion

What some have called the "Austin model of development" is, for the most part, just about bringing the state of Texas, through the university, into the local growth coalition as a leading actor. This changing political-economic dynamic helps explain why local power brokers from the business community moved toward the state and away from local government. Most authors attribute this shift to the rise of antigrowth forces in the local government and their unwillingness to support the business community (Farley and Glickman 1986; Orum 1987; Swearingen 2010; Kim 1998). Certainly this factor is important: members of the local business community would have tried to bypass their opponents in the local government by collaborating with actors at the state level, who were friendlier to their vision of Austin's future. Nevertheless, the evidence suggests that this relationship was also made possible by a change in the state's position toward the role public higher education could play in promoting regional growth. This new state involvement, in fact, proved to be a decisive factor in creating a new political space for an alliance among these actors, an alliance that would shape urban development in Austin. In other words, Austin's success in capturing the investment of MCC and Sematech was the result of a concerted state industrial strategy and the increasing willingness of government leaders to leverage state resources in a demand-side fashion to promote regional growth. As a consequence, the center of gravity for members of Austin's local growth coalition changed because state policy had empowered state actors over the local government.

These political shifts also transformed the role UT had in Austin's growth coalition. Up until the 1980s, UT's role in the growth coalition had been much more ancillary. Certainly by 1950 it was recognized that the university had made significant and positive contributions to the local economy beyond just being a major institution that contributed to private real estate development by bringing students and professors into the city. Instead, it assumed a role as the centerpiece of what would now be called the new knowledge economy. Research and science were increasingly valuable, and local growth advocates had seen the positive returns from integrating the university more into the local growth coalition and its efforts to bring companies to the region. Some scholars, therefore, see a common thread in Austin's historical evolution and claim that the presence of a prestigious research university that generates lots of employable workers and a creative atmosphere is the primary reason that many other firms ended up locating in Austin (Florida 2005, 79–81; Smilor, Gibson, and Kozmetsky 1989). While such claims may be overstated, and in some cases self-serving, such a narrative does recognize UT's central role in the development of the region's industrial high-technology infrastructure and, as a consequence, its increasing significance to Austin's growth coalition.

Yet such accounts are only partially correct because they largely fail to acknowledge UT's major role as the state's agent in land development ventures that were intended to secure external investments for Texas. It is therefore important to highlight the contrast between the two kinds of activities exercised by the university that increased the region's competitiveness: the first, supporting the production of engineers and interfacing with local companies to commercialize technologies; and the second, acting as a land developer that opened up a space for the growth of private corporations. The former action still engaged the university as a producer of knowledge, even if this production was subordinated to the interests of corporations, local business coalitions, and national (or state) competitiveness. The latter, however, opened up new possibilities by connecting a state university's endowment, and its bonding authority, to land development schemes designed to subsidize for-profit businesses and support the state government's industrial policy.

PART II

Urban Transformations

CHAPTER 4

Sustaining a Higher Quality of Life

The Environment, Crime, and the Remaking of Austin's Downtown

> The privileging of nature and natural processes, implemented through a city's environmental agenda, produces spatial, political and economic impacts for a group of people who, because of society's common perception that they have willingly rejected the formal economic system, carry the symbolism and reality of being cast as outlaws, outcasts, dangerous criminals or pitiful (and disdained) victims.
>
> **Sarah Dooling, *Ecological Gentrification: A Research Agenda Exploring Justice in the City***

Introduction

Austin has been one of the fastest-growing cities in America for the last twenty years. It has also acquired a reputation as one of the most progressive, creative, and tolerant cities in the United States. As Richard Florida, the guru for the creative class, notes, "The city consistently ranks among the top cities in the country in economic, recreation, and environmental listings. . . . Austin's success reflects its commitment to both high-technology industry and lifestyle amenities required to attract and to retain talent in the new economy" (Florida 2005, 80–81). Florida's narrative, commonly echoed by Austin's business elites, suggests that an enlightened ruling business class saw it to be in their long-term interest to embrace aspects of environmentalism in order to make Austin "the technopolis" of the Southwest (Kessler 2008). "While the business community and environmental community have had debates and disputes even fights over the years," said businessman Pike Powers, "I think all sides have come to realize in a mature fashion that protection of the environment and the enhanced insistence of water and air quality has made this a better place to live and a more competitive place to live" (Powers 2008).

Like other Sunbelt cities, Austin has grown dramatically since the 1950s. In 1940 the city had a population of about 90,000 with an incorporated area of about 30 square miles. By 1990 the population was about 470,000, a fivefold increase, and the incorporated area had increased sevenfold to about 225 square miles (Humphrey 2010). Regionally, the population in the 4200-square-mile area around and including Austin grew about 50 percent during the 1990s to about 1.25 million people in 2005 and is expected to double by 2025 (Steiner 2008). Although gross demographic increases are separate from the need to expand urban areas, the urban footprint of Austin has expanded because many people have settled in suburban and exurban communities that are connected by a vast network of federal and state highways. Fueled in part by private developers keen to take advantage of cheap land and a City Council whose main source of revenue is land taxes, this urban growth dynamic has dominated Austin since at least the 1960s.

However, since the early 1990s, Austin has earned a reputation as a very environmentally friendly city, with the local government taking a number of pro-

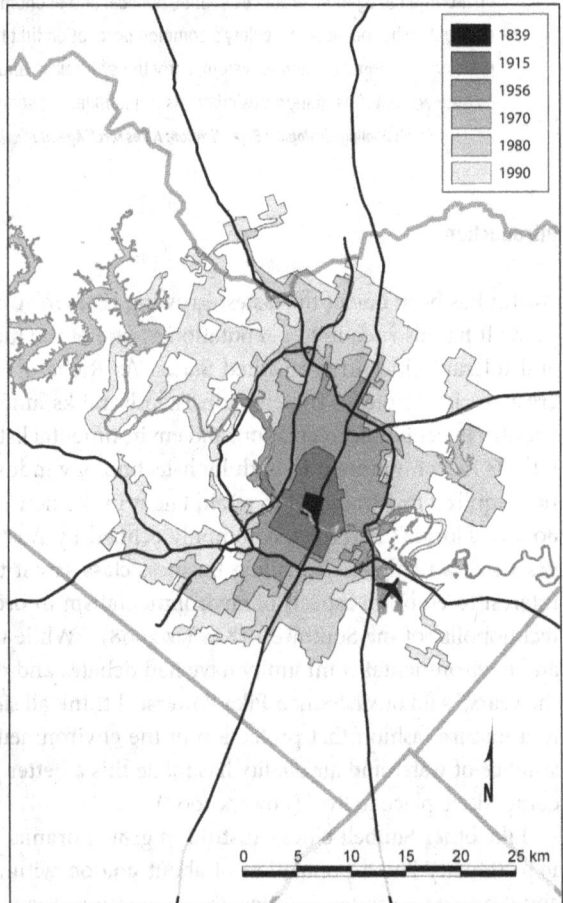

FIGURE 6.
The Growth of Austin City Limits. Based on map found in Robert James Ward (1998), "The Impact of Community Development Block Grant Funds on Neighborhood Revitalization: Austin, Texas." PhD diss., University of Texas at Austin.

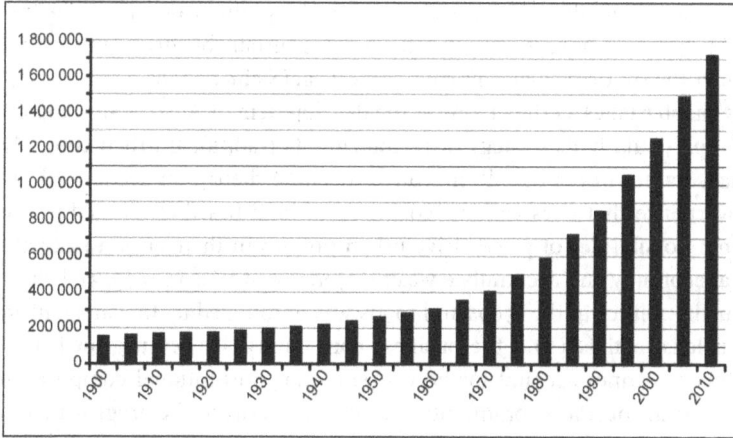

FIGURE 7. Austin MSA Population, 1900–2010.
Based on graph provided by Ryan Robertson, City of Austin Demographer.

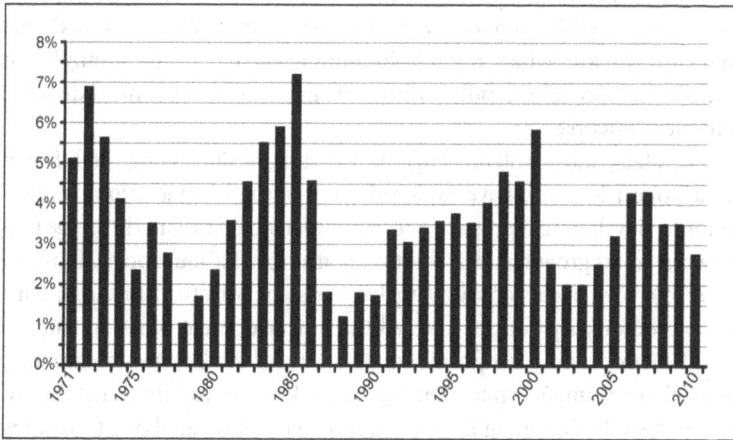

FIGURE 8. Austin MSA Annual Population Growth Rates, 1971–2010.
Based on graph provided by Ryan Robertson, City of Austin Demographer.

gressive steps to implement a range of innovative, ecologically sustainable initiatives (Gibbs and Krueger 2007). Since the development of the American Cities Business Journal's Green Cities Index, the city has consistently ranked near the top of the list (in 2010 Austin was number 4). It has even been described as the "Environmental City"—a distinction that is all the more surprising considering that Texas, according to a *Forbes Magazine* study, ranked thirty-fourth on the Green State Index in 2007 (Wingfield and Marcus 2007). While Austin is "the mid-sized American city with the most [urban] sprawl," its local government has attempted, more than its regional neighbors, to promote density in the traditional urban core in order to construct a more sustainable urban infrastructure under the rubric of environmental protection (Steiner 2008).

This chapter takes a critical look at the development of Austin's sustainability turn, particularly its strategy of revitalizing its traditional urban core. My analytical framework is based on growth machine theory, an account of urban governance that stresses how "coalitions of land-based elites, tied to the economic possibilities of place, drive urban politics in their quest to expand the local economy and accumulate wealth" (Jonas and Wilson 1999, 3). I show how Austin politics in the 1980s and 1990s was dominated by the antagonism between local activists and the business community. However, growth machine theory cannot account for how a temporary institutional compromise that favored the business community was reached between the progrowth and antigrowth coalitions. While, Jonas and Gibbs's concept of the "sustainability fix" helps solve this problem by adding urban competitiveness and entrepreneurialism to the equation and emphasizing how the making and unmaking of spaces creates preconditions for the reproduction of a system of accumulation, their theoretical insights do not capture the particular "sustainability fix" that has emerged in Austin, where the refashioning of the city center around ecological concerns worked in tandem with a program of social control directed against homeless people.

Other scholars, most notably Eugene McCann, Joshua Long, and David Gibbs and Robert Krueger, have explored some of the contradictions in Austin's economic development programs recently, focusing particularly on the tensions between smart growth and livability (Long 2010; McCann 2007; Gibbs and Krueger 2007). My conclusions are largely consistent with their findings, but I mainly explore how city policies affected the redevelopment of its downtown. Moreover, I focus on how the policies and actions of the business community and the local government, often working in tandem, moved from a strict law enforcement agenda ("cleaning it up") to a more environmental one ("greening it up"). I suggest that this change resulted from a political compromise between the progrowth and antigrowth coalitions because the latter were willing to accept the redistribution of the costs of growth away from "nonhuman species" and onto humans, particularly homeless people. The urban growth coalition certainly responded to a changing interurban competitive environment for investment that focused more and more on the traditional central business district, but the particular innovative policy solutions that emerged in Austin at that time, focusing on the revaluation of downtown, were born out of the struggle and political actions of local antigrowth coalition activists. These activists did not criticize how the policing of downtown was a precondition for the type of urban development that was being offered as part of the political compromise.

From the Growth Coalition to the Sustainability Fix and Social Control

In *Urban Fortunes* Molotch and Logan offer an account of why urban politics in a democratic society is dominated by a progrowth agenda, when in so many cases the costs of growth are borne by the majority of people while the benefits accrue to few (Logan and Molotch 1987). They argue that a small group of economic elites, particularly those whose fortunes depend on increases in the economic value of local real estate, has an asymmetrical influence on how cities are governed. Controlling a vast store of immobile capital tied up in investments in land and improvements on land, these local landowners try to capture revenue from more mobile forms of capital, which they siphon off in the form of rent. More intense land use permits a greater amount of rent to be taken from this "cosmopolitan" capital. City governments are compelled to increase growth for a number of reasons, most importantly because in the United States, local government tax revenues depend on increasing the appraisal or exchange value of land, and the intensification of land use means more tax revenue. Moreover, politicians' campaign funds are tied to the expanding fortunes of these local landed elites. While many individuals or groups object to the costs of growth, effective opposition, Molotch and Logan argue, is diffused by local government officials, private landowners, and the media, which present growth as good for everybody. The coalition of progrowth actors, both public and private, that organizes around these real estate interests forms the so-called growth machine.

Despite the growth machine's hegemony, opponents to growth do emerge. According to growth machine theory, groups such as environmentalists and neighborhood associations can pose significant challenges, in part because the costs of urban growth can be seen clearly in the degradation of a "good environment." While growth often happens at the expense of environmental quality (the encroachment into fragile animal habitats, increases in air pollution, degradation of water quality, and so on), the social environment (the loss of open space, demolition of historic buildings, traffic congestion, overcrowded schools, decaying neighborhood cohesiveness, and so on) is also critically important. Both environmentalists and neighborhood associations, according to Molotch and Logan, rely on a different conception of the value of land than the one put forth by urban boosters, one that focuses on investing in the quality of the "environment" over the economic return of investments. Instead of trying to increase the exchange value of land, these groups seek to preserve or enhance the use-value of their place.

While not alone in their belief in the detrimental social and ecological effects of urban growth, Molotch and Logan suppose that the urban political landscape is animated by the antagonism between progrowth and antigrowth coalitions, and, more importantly, that there could not be a temporary compromise that could help ensure the existence of the growth machine. Yet ecological and neighborhood concerns about "livability" and "environmental quality" seem to have become central to a growth machine agenda. More and more cities present themselves as more "environmentally friendly," "sustainable," or "smart," and local development policies at least partially seek to internalize the social and ecological costs of growth (Krueger and Gibbs 2007; Portney 2003; Whitehead 2010; Moore 2007). Some authors have seen these new policies of sustainable growth as merely an ideological smokescreen that repackages the old "growth machine" into a new "smarter growth machine" and replaces the rhetoric of "value-free" growth with "sustainability" while generating the same negative social and ecological results (Gearin 2004; Troutman 2004).

In contrast, other scholars suggest that the recent fusion of urban growth and sustainability in urban politics is a result of the substantial reworking of local urban political structures within the system of interurban competition (Keil and Whitehead 2012; Dierwechter 2008). For instance, the focus on the environment can serve two purposes: it allows for spaces in the city to be cleared for new rounds of development, and it diffuses potentially disruptive political opposition (While, Jonas, and Gibbs 2004, 554). The production and destruction of urban landscapes is a precondition for new rounds of capital accumulation and expansion. For instance, factories that were built as investments and that produced an industrialized landscape were later unmade by deindustrialization, only to be remade into lofts or retail areas or demolished for some other purpose. To shape how their cities are remade, city governments increasingly take a very entrepreneurial stance and leverage local resources or amenities to lure mobile capital or foster local economic activity by providing incentives for private investment (Savitch and Kantor 2002). Assets—such as access to ports, cheap labor, and cultural amenities—increase competitive advantage and enhance a city's marketability by promoting it to private investors (Hall and Hubbard 1998; Harvey 2001; Leitner 1990; Tretter 2009; Cochrane 2007; Hackworth 2007).

Environmental amenities are yet another asset that can be used to boost a city's fortunes, and the present system of urban entrepreneurialism may depend "on the active remaking of urban environments and ecologies. . . . [A]ctive environmental policies and interventions have . . . been important in opening up actual urban spaces for new waves of investment" (While, Jonas, and Gibbs 2004, 550). The incorporation of ecological concerns into an urban growth agenda may have a significant impact on urban governance because it helps establish a framework for a temporary institutional compromise among various interest groups, which While, Jonas, and Gibbs call the "sustainability fix." They contend that urban elites have been forced to address the environment as a part of the growth agenda because of increasing political awareness about global environmental degradation, middle-class concern for a higher quality of life, and image campaigns that represent these old industrial cities as clean. For them, the "sustainability fix" relies on the selective incorporation of environmental concerns by urban managers, primarily in postindustrial cities, which are trying to negotiate a tenuous political place between ecological concerns and the economic realities of global economic urban restructuring.

Urban sustainability is also more and more connected to the securing and policing of city centers (Helms 2008, 15). Issues of personal safety and security became linked with urban "livability" and "quality of life" during the 1990s, and addressing crime became a significant benchmark for a city's competitiveness (Coleman 2003; Oc and Tiesdell 1997). To address real and perceived threats to public safety, cities throughout the United States began to pass local ordinances that would prosecute "quality of life offenses" or minor "victimless" misdemeanors, a process sometimes called "zero-tolerance policing" (Belina and Helms 2003). The idea developed in part out of the "broken windows" theory of crime, as popularized by George Kelling and James Wilson, which uses the metaphor of a broken window to describe how small offenses lead to large social problems. The theory suggests that if small amounts of social disorder, like broken windows, remained unfixed, a general atmosphere of decay and lawlessness will be the result (Herbert and Brown 2006). Therefore, strict and nondiscretionary police action should be taken against all minor crimes and offenses such as drunkenness, panhandling, vagrancy, and failure to pay for public transportation, and these crimes should be punished to the maximum extent permitted by the law.

In many cases, the enforcement of these crimes would selectively target certain areas of a city, particularly its downtown, and certain people, especially the homeless. Having people sleeping on streets, panhandling, and loitering, scholars and public opinion makers argued, created an atmosphere of decline and a vicious circle, where investment driven away by fear led to the further decay of city centers and encouraged the arrival of more homeless people (Mitchell 1997). To halt this decline and improve the image of a city's downtown, urban revitalization efforts increasingly combined law enforcement with other economic incentives. Making the city secure, safe, and clean would encourage "regular," that is, middle-class, people to shop, work, play, live, and even buy real estate, such as condominiums. The goal was to create a twenty-four-hour downtown that "weaves together upscale retail and world-class activities with gleaming office towers and affluent urban residents" (Gibson 2004, 2). Increasingly the numerous ecological and sustainability benefits that come from a revitalized and active downtown, particularly in terms of reduced carbon emissions from driving and energy savings from buildings, have taken priority over crime prevention as the leading indicator of the "quality of life" in a city. But, as Dooling has argued, ecological benefits have also become one of the main justifications for further excluding homeless people from city centers (Dooling 2009).

Austin's Growth Coalition and Its Opposition

During most of the twentieth century, the hegemony of the business community in Austin politics was unquestioned, and the primary vehicle that expressed business interests was the Austin Chamber of Commerce (Bohmfalk 1968, 10). Many of Austin's mayors and City Council members have served on the board of directors of the Chamber of Commerce, sometimes as its chair or executive director (Orum 1987, 56). In the early portion of the twentieth century, the Chamber of Commerce's efforts focused primarily on industrializing Austin. Despite its efforts, Austin did not become a major industrial and manufacturing city, and because most of the land in the city did not produce tax revenue (as it was state land), the Chamber soon began to focus on "the improvement and enlargement of its educational, governmental, recreational, and commercial facilities" (Bohmfalk 1968, 24). With the emergence of a technological sector in the late 1950s, industrialization once again became a possibility. In 1955 the business community had the local government "entrust creation of the Austin Development Plan to a subsidiary group of its [the business community's] own choosing" (Moore 2007, 36). The development plan, discussed in more detail in chapter 6, was eventually adopted in 1961, allowing the city government to offer incentives to attract "footloose" businesses. Its effects were first realized when, after a noteworthy multiyear marketing campaign by the Chamber of Commerce, several national and international companies located in Austin's suburbs (Orum 1987, 245–248). By the 1980s, as we have seen, the growth of both national and local technology firms in Austin was substantial; the city had become what a University of Texas at Austin business professor called a "technopolis," and the Chamber of Commerce is largely credited as the institution that made this happen (Kessler 2008; Arbingast and Ryan 1982).

Two reports published about a decade apart by the Chamber of Commerce reveal an increasing concern by the business community with "quality of life" issues, particularly environmental concerns, and their impact on Austin's competitiveness. "Creating an Opportunity Economy," published in 1985, had a substantial influence on local private and public policy and, to a large extent, became the long-range competitive plan for the city (Engelking 1999). Focusing on ways to expand Austin's technological sector, the report outlined key issues of public investment and infrastructure that the Chamber of Commerce believed needed to be addressed to ensure a robust period of economic growth, and it stressed the importance of maintaining a high "quality of life" standard for Austin's future economic success. The city's competitiveness, the report noted, "depends as much on a community's quality of life as on its work force. Appropriate education and training are as necessary as adequate physical infrastructure" (Austin Chamber of Commerce 1985, 5). While the report briefly mentions concerns about the negative impacts of the burgeoning technology sector on the quality of Austin's social and physical environments, it does not propose any solutions. Just over ten years later, in 1998, the Chamber of Commerce offered a new plan in a report titled "Next Century Economy" that was more explicit about the role of social and environmental infrastructure in the city's competitiveness. "Given Austin's economic direction," the report said, "environmental and social issues are important because they are also critical inputs to its long-term economic competitiveness" (Lyman 1998, 5). While antigrowth coalition members suggested that with the release of this report the Chamber of Commerce had come to recognize "the environment" "as important," the only environmental issue the report mentions at length is traffic congestion (Arnold 2008). Like the 1985 report, the 1998 plan's emphasis on quality of life, particularly on environmental concerns, was important only in terms of promoting Austin's ability to attract and retain new highly mobile, high-technology employees.

Despite the reticence of the two reports on the subject of environmental quality, by the early 1980s people were beginning to suggest that the growth of the high-technology sector and the fast growth of Austin's economy were causing significant environmental degradation (Myers 1984, 22–24). High-technology firms often located far from Austin's downtown on the city's edges, and as highly educated and wealthy employees began moving near these firms, the city's urban footprint expanded drastically. Moreover, and perhaps more importantly for the city's environmental quality, high-technology firms and wealthier residents seemed, in many cases, to prefer locating in the karst environment of the Edwards Plateau, a large inland limestone uplift in the western part of the city. Spanning much of the western part of Austin, the plateau forms the sources for the Edwards Aquifer, a series of underground waterways that provide much of the region's fresh water. Up until the 1990s, the area was sparsely settled because it is fire prone and its soil cannot sustain large-scale agriculture. Increasingly valued for its scenic beauty, the area has a limestone bedrock that is very accommodating to development because it does not have the same problems of subsidence and flooding that are common in Austin's eastern areas (Knight 2008; Tretter and Adams 2012).

As early as the late 1970s, Austin's rapidly expanding urban sprawl had begun to generate a significant political response, with two factors converging into an inchoate antigrowth coalition: the rise of environmentalism as a political movement and an increasing awareness of the costs of growth, particularly in wealthier and predominantly white neighborhoods on Austin's west side (Swearingen 2010). On the one hand, environmentalism was emerging nationally as an independent political movement that resonated with many Austinites who were witnessing the urbanization on the Edwards Plateau and the resulting threats to its ecological stability. On the other hand, urban sprawl was contributing to the decline of older, wealthier west side neighborhoods (where many of these environmentalists lived) and the commercial vitality of downtown (Orum 1987,

290). To add insult to injury, the city and state highway authorities were build-ing a new highway, which connected these newer suburban developments to the city's center, straight through the heart of Austin's west side.

By the mid-1980s progrowth and antigrowth coalitions formed the two pri-mary political factions in the city, and there was apparently little room for com-promise (Smilor, Gibson, and Kozmetsky 1989, 55). The main battle was over restrictions on impervious surface cover over the Edwards Plateau, not over suburban growth (Cooksey 2008). The antigrowth coalition won its first sub-stantial legislative victory in 1985 when it took control of the City Council. Led by Mayor Frank Cooksey, the council attempted to minimize potential environ-mental degradation over the Edwards Plateau by regulating the quality of water in natural waterways that formed the Edwards Aquifer. After years of discussion and opposition from the business community, the Cooksey Council eventually passed a local water ordinance, which, despite being considered weak by many environmentalists, did offer regulatory protection for some waterways around the city (Swearingen 2010, 142–147). However, in 1988 the antigrowth coalition suffered electoral losses, and a new mayor, Lee Cooke (the former chief execu-tive for the Chamber of Commerce from 1983 to 1987), and several other more growth-coalition-friendly candidates won seats on the council. The issue of de-veloping over the Edwards Aquifer once again came to a head in 1989 when a land speculator who had bought land along one of the aquifer's main tributaries proposed building a massive new project requiring the council's approval. Fol-lowing public outrage at the request, the council, surprisingly, failed to approve the proposal and passed a short-term moratorium on new development until a more comprehensive water ordinance could be written. In response, progrowth forces, led by the Chamber of Commerce, financed a slate of candidates that won in 1990 City Council elections. Much friendlier to the business commu-nity and suburban developers, these members of the council were openly hostile to environmental and neighborhood concerns and resisted enacting a stronger comprehensive urban watershed ordinance (Swearingen 2010, 151–154; Slusher 2008).

In the early 1990s, with neither side having a clear majority, the City Coun-cil became locked in a long-term political stalemate that slightly favored a pro-growth agenda. Antigrowth coalition groups decided in 1991 to organize a peti-tion drive to put its own water quality ordinance (known as the sos ordinance), stronger than any previous water ordinance passed by the City Council, on the ballot for a public referendum. Fiercely opposed by residential developers and the Chamber of Commerce, the ballot measure passed, in 1992, by a margin of about 2–1. Although machinations before the referendum vote by progrowth members of the City Council and subsequent interventions by the state legisla-ture after its passage stymied the ordinance's overall effect in limiting impervious surface cover over the Edwards Aquifer, the overwhelming popular support for the ordinance ensured that environmentalists' demands could no longer be ignored (Bunch 2008; Swearingen 2010, 179–180).

The Downtown Austin Alliance: From Cleaning Up to Greening Up

While the political fight between the progrowth and antigrowth coalitions was raging over water quality, property rights, and the future of Austin's suburbs, the transformation of the city's downtown remained largely outside the discussion. In the early 1980s, there was a substantial property boom in downtown Austin, and for the first time large modern glass buildings became a part of the city's skyline. By the end of the decade, however, Austin's building boom had become a financial bust; many recently built large buildings were boarded up, and there was a glut of vacant office space in the city's downtown (Barna 1992).

In 1987, the Chamber of Commerce, partly in response to the downtown real estate crisis, began to focus on "the revitalization of downtown" Austin (Powers 2008). The Greater Austin Chamber of Commerce's Downtown Revitalization Steering Committee and the Downtown Commission (an advisory board that works with the City Council's Downtown Subcommittee on issues related to development in the downtown area) urged the City Council to fund an evaluation of the area by a Regional/Urban Design Assistance Team (R/UDAT).* According to community and environmental activists, efforts by the R/UDAT and the Chamber of Commerce to focus on revitalizing Austin's downtown had a "good response" in the antigrowth coalition, although it remained a rather marginal concern compared to water quality (Arnold 2008).

One of the R/UDAT reports, written in 1992, offered a detailed revitalization program, proposing recommendations in the following areas: urban design, the natural environment, community issues, cultural arts, transportation, and economic development. "Downtown," the report concluded, "must regain its position as the economic hub of the city through the implementation of a progressive economic development plan. Downtown, however, is no longer perceived solely as a center of commerce but must also incorporate culture, entertainment, tourism and residential neighborhoods to maximize its chances for prosperity" (City of Austin 1992, 97). Significantly, the report also suggested establishing a private-public partnership that would coordinate the development agenda for

---

*An R/UDAT is an extensive urban evaluation undertaken by professional architects and planners on urban design and planning issues. The experts visit a city and conduct several days of intensive workshops with community stakeholders, and then within a year return with an implementation program.

downtown, an organization it called the Downtown Management Organization (DMO). The report noted:

> Acting as Austin's Downtown advocate, a DMO is management of Downtown by Downtown. The DMO will: provide leadership and oversee promotional, management and financial concerns; be a unified voice for property owners, local government, tenants and Downtown patrons; represent Downtown at the local, state and federal levels; ensure accountability in planning, funding and implementation of projects and programs; produce additional funding to implement high-priority projects and programs; interact with social, business and adjoining neighborhood organizations; underscore Downtown's role as the seat of government and the economic, cultural and social focus of the city; and consolidate the strengths of Austin's business and civic leaders. (City of Austin 1992, 98)

In 1993 the City Council voted to create the DMO, which was renamed the Downtown Austin Alliance (DAA) in 1995 (City of Austin 1993). The original organization was composed mainly of businesspeople and property owners in the downtown area, and while local governments contributed funds to the DAA, its primary financing came from a special tax assessment on privately owned properties within the downtown (Smith 1994).

The major factor limiting economic redevelopment efforts in the downtown area was the problem of "safety and crime," and the DAA's central mission was to "improve downtown security" and beautify the area (Todd 2008; Smith 1994; Schwartz 1996b; Windle 1994; Barnes and Warren 1995). While some public works projects were necessary to clean up downtown, the project also involved a significant law enforcement component and the establishment of a zero-tolerance policing (ZTP) zone. Beginning in 1994, about two years after the alliance was formed, the organization began its most successful project, the Downtown Rangers, a uniformed civilian force established to police quality-of-life offenses such as "panhandling, drunkenness, car theft, and similar crimes" (Beal 2001, City of Austin 1993). It was also partly funded by the city. Jose Martinez, executive director of the DAA, said, "We surveyed people downtown and found, not surprisingly, that their No. 1 concern was crime. We [the DAA] want the attitude of zero tolerance to be the attitude of downtown" (Banta 1994). Unlike in many other city community policing initiatives, the Rangers were not police officers. They were a private security force that could not arrest or ticket people but that were the "eyes and ears" of the police, reporting crime and acting as "Austin's ambassadors" by assisting tourists and visitors in the central business district.

Cleaning up downtown also involved removing certain kinds of people, namely the homeless. During the 1980s, the homeless population in Austin had grown significantly, and Snow estimated their numbers to be between about one thousand and thirteen hundred (Snow and Anderson 1993, 19–20). Since nearly

all the services for the homeless were in the central business district, they had
become a fixture of the city's downtown area. In the 1990s the homeless were
identified as the major obstacle to growth and the reason why downtown Austin
was so poorly maintained; they became the main targets of the enforcement of
ZTP "quality-of-life" crimes. Additionally, the DAA supported a number of punitive measures specifically targeting the homeless population, such as asking City
Council to delay the reappointment of judges thought to be lax on sentencing
for "nuisance crimes" and lobbying the City Council to pass a camping ban that
imposed a five-hundred-dollar fine on people sleeping or attempting to sleep
in public (Austin American-Statesman 1995b; Dworin and South 1996; Lindell
1996). The passage of the camping ban was called a "moral victory" by Jose Martinez, who went on to say, "We need the ban to make it safe and clean here and
to make the downtown attractive" (Dworin 1996).

By 1996 the DAA's efforts, along with those of its security arm, the Downtown
Rangers, had proven so successful that Martinez, by then the outgoing executive
director of the DAA, would proclaim that the group was moving "beyond clean
and safe" (Schwartz 1996a). Six months later, the newly appointed executive director, Charlie Betts, suggested what this meant: "Though the term 'development' usually does not bring a smile to the face of Austin's environmentalists .
. . downtown development should" (Restrepo 1996). The "development" Betts
was alluding to was a high-end, large-scale residential building that had been
made possible by the 1995 change to the land development code to allow the
construction of homes in the downtown area. Now that downtown was "secure
and clean," a new phase in development could begin that involved creating a "24
hour downtown" (Greenberger 1997b; Tyson 1995).

Smart Growth and the Greening Up of Downtown

The Downtown Austin Alliance's suggestion that downtown revitalization
should be a priority for environmentalists came at a turning point in how the
business community was responding to the antigrowth coalition. In response to
the business community's opposition to the Save Our Springs (SOS) referendum,
in 1997 the antigrowth coalition successfully got nearly all their preferred candidates elected to the City Council, which became known as the "Green Council"
(Zacharias 1997). Following this notable electoral defeat, the Chamber of Commerce, while continuing to support exurban growth, was much more willing
to cooperate and eventually came to embrace a version of environmentalism
that, as Betts had suggested, would place a premium on the redevelopment of
downtown.

The phrase that encapsulated the new City Council's agenda was "smart
growth." Already having a great deal of currency in urban planning, smart

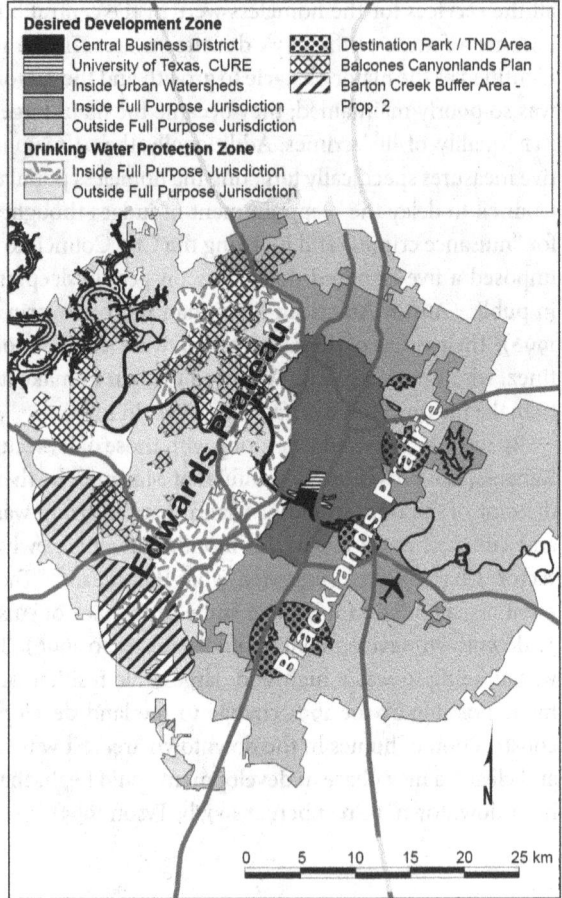

FIGURE 9.
City of Austin's Smart Growth Zones.

growth, as Dierwechter has argued, should be understood as a policy that developed in the long history of local and state government attempts at growth management but one that focuses on free-market rather than regulatory actions to contain growth (Dierwechter 2008, 36). Echoing President Clinton's Council on Sustainable Development's "vision of a life-sustaining Earth," the vision of smart growth presented in Austin suggested that there could be a balance between economic prosperity, ecological preservation, and social equality (President's Council on Sustainable Development 1996). The major proponent of smart growth was Austin's newly elected mayor, Kirk Watson, who had "used the language in his election campaign" (Arnold 2008). "Austin's government had been dominated for too long by win-lose, all-or-nothing politics pitting the environment versus economic development," Watson wrote. "Not only had the every-day interests of Austinites been subsumed by these battles, but the warriors had lost sight of all the common ground they shared and were failing to work together to safeguard the city's future" (Watson 2010). Watson had been both a chairperson of the Chamber of Commerce and an active member of the Texas Democratic Party, and it is interesting to note that his legacy is claimed by both pro- and antigrowth coalition members (Bunch 2008; Powers 2008). Most important, however, are the terms of the compromise offered to resolve the conflicts between the two coalitions.

While equity was a part of the smart growth plan—over the DAA's objections, the City Council voted to build a homeless shelter in the central business district called the Austin Resource Center for the Homeless (ARCH), and some of the city's minority populations did receive some benefits—the essence of the compromise lay in its ability to steer growth off the Edwards Plateau and closer to the city's central business district (Ball 2003; Osborne and Scheibal 2003). A map released by the City Council showed the city divided into three different development zones: a drinking water zone, an urban desired development zone, and a desired development zone (Clark-Madison 1998). "Keep [the area over the Edwards Aquifer] as low density and protected as we can by steering our more intensive urban development to the east and downstream of the Edwards aquifer," Bill Bunch, a leader of SOS, noted. This idea was "restyled during the green Watson council as the desired development zone" (Bunch 2008). Infill and higher densities would be pushed into the desired development zones, particularly the urban desired development zone, which covered all of the traditional central business district.

These designations were not legally enforceable, as the Council lacked regulatory authority; instead, tax incentives or subsidies were offered to developers and corporations as an enticement to support the plan (McCann 2003, 168). Two bonds, presented in a package in 1997, became the hallmark of the city's smart growth strategy to revitalize the downtown. The first bond authorized raising the city's hotel room tax to 15 percent to finance the expansion of the Austin Convention Center and the construction of a flood remediation project, which was designed to reclaim a large portion of land for redevelopment. The other bond authorized $65 million to buy 15,000 acres of undeveloped land over the Edwards Aquifer through a water/wastewater utility surcharge. Both bonds passed and were strongly supported by key institutions and persons in both the pro- and antigrowth coalitions. For instance, the SOS Alliance campaigned vigorously for the first bond, and in return the Chamber of Commerce backed the land-acquisition proposal, which barely passed and probably would not have without the business community's support (Lindell 1998).

Austin's smart growth plan emphasized the redevelopment of the city's business district and helped solidify the area's growth potential by permitting more intensive land use. Pushing more aggressively for the use of local government funds and regulatory easements to ensure the building of more residential and commercial projects in the city's downtown, the DAA began to promote the idea that Austin's city center was a truly twenty-four-hour downtown and should include more housing, in addition to restaurants and other forms of entertainment (Breyer 1997). The fact that Austin's downtown had already been home to many residents, namely homeless people, remained virtually unmentioned in the debates between the rival factions. However, the language of environmental stewardship and the support of the antigrowth coalition were central to the fulfillment of this vision, and in 2000, the DAA even gave SOS one of its yearly "Impact Awards." The then-outgoing DAA chairman (and soon to be elected mayor) Will Wynn said, "They [SOS] helped champion downtown as an alternative to environmental degradation and continued suburban sprawl" (Lindell 2000). However, as the evidence I have presented suggests, SOS was not championing downtown as an alternative to development in the suburbs. These ideas came from the business community and became part of a compromise offered to the antigrowth coalition.

Conclusion

Some scholars have attempted to explain the tidal shift in urban policy toward programs like Austin's smart growth initiative, sometimes called "sustainable urbanism," by appealing to growth machine theory because it offers an explanation for why growth machines support intensification of land use in city centers: taller buildings and intensification of land use means more rent and therefore more tax revenue for local governments. Smart growth, these authors contend, still supports the growth machine agenda but in a different guise. Troutman, for instance, notes, "Smart growth reduced the problem of growth to one question: where to put it. Urbanized communities were good: greenfields were bad" (Troutman 2004, 617). Questions about the environmental costs of all development and the societal relations that give rise to unsustainable living are excluded from the discussion, and, as the Austin case study shows, environmentalists and neighborhood groups are expected to support bigger, more compact, and more intensive development in city centers because it is intended to divert development away from suburban areas. While some proponents of smart growth do mention concerns about its negative effects on the affordability of new housing and retail outlets, the ecological benefits are nearly always unquestioned. It is assumed, by people like Edward Glaser, that cities are "greener" than elsewhere because people use their cars less often and have a much smaller carbon footprint when they reside in more compact and higher-density areas (Glaeser 2009). However, this claim is suspect because there is little evidence offered to suggest that carbon reduction is the best metric for evaluating environmental stewardship and long-term ecological and social sustainability. More importantly, it is also unclear whether compact cities do, in fact, have a smaller ecological footprint (Neuman 2005). Instead, smart growth is very compatible with a traditional growth machine agenda because it does not seek to alter any of the fundamental social relations that produce ecologically unsustainable urban growth. Smart growth becomes much more about creating an environment of sustainable returns on investments than about environmental stewardship, even if there might be some ecological benefits.

Growth machine theory has limited application in the case of Austin because it cannot account for the compromise that was forged between the progrowth and antigrowth coalitions (a compromise that helped sustain the growth coalition's agenda) because the theory supposes only two outcomes: a political stalemate or the victory of one of these two coalitions. Instead I have shown how the antigrowth coalition did not undermine a growth agenda but instead made significant contributions to creating the conditions for new rounds of profitable capital accumulation. Austin's smart growth plans made the city more competitive because in order to attract what Florida has called the "creative class," certain infrastructural preconditions had to be met, particularly in regard to the preservation of open space. The progrowth coalition was quick to learn that local ecological amenities with no apparent economic value could be harnessed for the city's growth. However, the more substantial change, which would have a more significant impact on the redevelopment of the city's central business district and the sustainability fix, concerned the push to create a compact twenty-four-hour city complete with taller, more land-intensive buildings for residential, commercial, and retail purposes. The actions of antigrowth coalition activists, with their concerns about the impact of urban development on the region's ecology, became the centerpiece of an entrepreneurial strategy that has helped reposition downtown Austin and the city as a whole in the competitive hierarchy for investment.

Focusing on the institutional, or the "sustainability fix," is not sufficient to account for the political compromise that emerged in Austin. The fix was not only institutional but spatial. The formation of the DAA and the legislative actions taken by Austin's City Council were a significant part of the institutional compromise that developed between the antigrowth coalition and the business community, but the compromise was preserved by the shift in where development occurred, which affected who and what bore the costs of growth. The antigrowth coalition could ultimately only deflect the costs of growth from one "victim," the natural environment for endangered species, to another, the urban environment of the homeless. For the program to be successful, it had to be married to a significant law enforcement component that could support a particular vision of an enhanced "quality of life."

CHAPTER 5

Contesting Sustainability

"Smart Growth" and the
Redevelopment of Austin's East Side

> What is an environmental group? Who is an environmentalist? How might different
> kinds of environmental groups influence the state of the environment—and the state
> of society? These charged questions lie at the heart of the problem of how to define
> environmentalism and its future direction.
>
> **Robert Gottlieb, *Forcing the Spring: The Transformation***
> ***of the American Environmental Movement***

Over the last twenty years, two processes have emerged as the dominant forces shaping the politics of urban development and planning practice in cities throughout the United States: the development of a more competitive international environment for investment and an increasing recognition that reorganizing cities is a necessary part of the solution to current environmental problems. Taken together, these two processes have significantly restructured the organization of urban space, particularly traditional central business districts. While distinct, the two are also interrelated. It is impossible to grapple with the myriad of global environmental problems without understanding the impact of patterns of investment and disinvestment and the environmental limits of growth.

On the one hand, competition among cities for investment has become more pronounced as revolutions in telecommunications and transportation have annihilated the distances between cities and reduced the friction of moving goods, information, and people across space. People and places are more interconnected, and local differences often seem to be eroding because in many respects the world is more uniform and, figuratively, smaller. This is not to say that the world's geography is even. The relative advantages of a location's distinctiveness have not diminished but are actually becoming more pronounced. To distinguish themselves from their competitors and to boost their positions in the hierarchy of the global investment marketplace, some cities have been very effective at entrepreneurially using their local resources (Hall and Hubbard

1998; Harvey 1989a; Savitch and Kantor 2002; Tretter 2009). The cultivation of environmental amenities and an ecological image has become part and parcel of a strategy to spur economic development because it has been shown to attract and retain educated and highly skilled workers and promote industrial development in innovative sectors such as renewable energy (Dooling 2009; While, Jonas, and Gibbs 2004; Jonas and While 2007).

On the other hand, while environmentalists once criticized cities as unnatural and antithetical to an ecologically sustainable future, cities are increasingly understood to be a necessary part of the solution to global environmental problems. Whether the issue is global climate change, fresh water, or open habitat, environmentalists have come to realize that reorganizing cities offers a unique opportunity to make societal relations with nature more sustainable (Dittmar 2009; Keil 2007). There is a growing recognition that solving tomorrow's environmental challenges requires a more holistic vision of the urban and nonurban today: that cities can be refashioned, becoming more sensitive to potential environmental impacts and integrated with natural surroundings. Although some environmentalists, such as Alan Weisman, have proffered a sustainable vision of a future world without humans (Weisman 2007), others focus on ways to avoid this bleak prospect for the collective human future. As more than 50 percent of the world's population currently lives in cities and makes taxing demands on the physical environment, many environmentalists insist that avoiding such a future requires alternative urban structures and processes.

Sustainability has emerged as a concept around which a comprehensive framework has developed to address a range of urban problems. It has become the dominant regulating principle in the formation of urban policy and planning theory. Deeply intertwined with the notion of sustainable development introduced by the Brundtland Commission, which stated that societies should meet present needs without compromising the ability of future generations to meet their own needs, sustainability policies have come to encapsulate a wide range of urban-planning practices, from bike lanes to LEED-certified buildings and affordable housing (Whitehead 2007). While there is not a consensus about what practices and policies are "really" sustainable, it is fair to say that sustainability is presented as an ideal, where people strive for a kind of development that does not create net social, economic, or ecological losses.

Although the idea of sustainability often is used to characterize a range of economic, social, and ecological problems and processes, from public health to new technologies to tourism, that mean more than just environmental awareness, much of the growth of its popularity can be attributed to the mainstreaming of environmentalism. In a popular context, sustainability and better environmental management have become synonymous, a pairing that has often led to the erasure of social and economic concerns elsewhere articulated as part of the sustainability ideal. For instance, in the work of architect Douglas Farr, "sustainability" simply means recognizing the importance of the environmental limits of growth and creating urban environments that provide a sustainable return on financial investments such as real estate (Farr 2008). But while it may be possible to create temporary and profitable environmental sustainability fixes for urban development, these fixes are often detrimental to social equality, similar to how some urban industrial and land-use policies, instituted to provide job opportunities and improve social welfare, had drastically negative environmental effects (Campbell 1996).

Since the early 1990s, as chapter 4 discussed, Austin's local government has attained some notoriety because of its commitment to ecological sustainability goals. The city has implemented a number of initiatives that have increasingly figured as central parts of the city's strategic planning (Lyman 1998). As the previous chapters have stressed, Austin's highly technically skilled workforce, the state government, and the University of Texas at Austin have long been recognized as distinctive assets, but increasingly the city's elites understand the city's compactness, political concern with the environment, and ecological amenities as locational advantages that can increase the competitiveness of Austin's central business district over its regional and national rivals in the struggle for investment (McCann 2007, 189). In a sense, the principles of developing an environmentally sustainable city are underwriting the investment campaign because the rationale for reorganizing the city's landscape (investing in Austin's urban core and redeveloping its downtown) is that doing so is environmentally friendly and increases Austin's competitive stature. Hence, it is increasingly apparent that a version of environmentalism has become inseparable from, compatible with, and even beneficial to the city's fortunes. Despite the fact that a more inclusive definition of sustainability often involves more than just ecological concerns, the dominant vision in Austin is organized around a specific idea of environmentalism that is informed by the political conflicts that emerged over the development of the city's western suburbs and is controlled by people who are mostly white and have accumulated significant economic and cultural capital.

In the literature on urban environmental politics, many authors point out a substantial split between "environmental sustainability" groups and what Agyeman has called "just sustainability" groups (Agyeman 2005, 39). The former tend to be more narrowly focused on environmental issues such as habitat preservation, open space, and energy consumption, while the latter promote a broader "civic environmentalism" that uses a different definition of the term "environment"—one that includes people and where they live, work, and play as well as the "physical and natural worlds" (Bullard 2007, 10). There is a core difference between these two visions. The former wants to extend the sphere of justice to include nonhuman species and seeks to create such mechanisms as regulations, incentives, and technological improvements to internalize (that is, to more fully incorporate and account for the negative life-cycle effects of any good or service that is produced) the costs of human modifications to the natural world. In contrast, the latter in principle favors extending these concerns to nonhuman species but also conceptualizes environmental issues as a matter of civil rights; this vision explicitly ties ecological concerns to social problems and turns "environmentalism" into an issue of social justice. The problem is that the latter vision calls for internalizing the social costs of growth, which can work at cross-purposes to achieving ecological goals. In Austin these two visions of "nature" are held by people from different parts of the city, and while the split between the two has been overcome many times, for instance in addressing the impact of industrial effluent on nonwhite communities, it has been a significant source of friction, particularly in relation to policies aimed at the redevelopment of the city's downtown.

While both the "environmental" and "just" sustainability visions offer valuable contributions to thinking about how to construct alternative forms of urbanization, only a selective environmental sustainability framework has been effectively incorporated into the hegemonic vision of Austin's strategic growth plan. While many environmentalists lament that subsuming their pro-ecological ideals into the business community's progrowth agenda is a form of co-optation that has limited more radical transformations, politically it is important to consider how such a selective incorporation of environmentalism was possible. My argument is that the answer can be understood only by asking what environmentalists counted as "the environment." I will show how the present political arrangements between environmentalists and the business community were born out of a compromise stipulating that environmentalists would support the remaking of Austin's downtown if the business community supported preserving an area in the suburbs for nonhuman species. The business community thought the compromise would help boost Austin's competitive edge because preserving a section of the "pristine environment" in the suburbs would increase Austin's ability to retain and lure highly educated workers (Preston 1998). In addition, the redevelopment potential of downtown was so great that it would offset any prospective downside of limiting growth elsewhere. Environmentalists were satisfied because spatially, the agreement meant limiting development in a fragile ecosystem and promoting future growth closer to downtown rather than in the suburbs. The unstated assumption of both sides was that the transformation of downtown's physical and social environment was not an environmental problem. An invisible boundary circumscribed the limits of what was considered an environmental concern. Environmental issues focused only on internalizing the effects of urbanization on nonhuman species; environmentalists were not concerned about the externalization of the downtown redevelopment's environmental costs onto people (both homeless people and communities of color).

Austin's Two Environmentalisms

On March 31, 2010, I went to the release party for a book written by Scott Swearingen, an urban sociologist who teaches at St. Edwards University in Austin. Scott is an acquaintance, and he had given me an advance copy of his book, *Environmental City: People, Place, and the Meaning of Modern Austin*, which I had read about a year prior, so I went to congratulate him.

The event was held in a reserved section at Threadgill's, a local restaurant chain famous for its southern comfort food and live music. Located along Riverside Drive in West Austin, the restaurant is very close to where the Armadillo World Headquarters, one of Austin's most significant live music venues, once stood. The event was hosted by the Save Our Springs Alliance (SOS), which widely publicized the event. Founded in 1990 as a coalition of groups and individuals contesting a proposed development in the Barton Creek watershed, SOS's main concern continues to be the potential damage to water quality caused by urbanization and the loss of natural habitat over the Edwards Aquifer. Widely considered Austin's most powerful environmental organization, SOS has won several significant state court cases and was victorious in getting the Barton Creek salamander (*Euryce sosorum*, named after the group) listed as a federally endangered species.

The event was first and foremost a book signing, but it was also a celebration. Several attendees told me that it was a reunion of sorts for the entire environmental community, and people who had not seen each other in many years were catching up; most knew Swearingen, who was deeply involved with Austin's environmental community during the 1980s, and his book tells their story. It is a detailed case study of how disparate and nascent local neighborhood and environmental activists became a unified and strong political movement. It contains personal stories, photos, and many invaluable firsthand accounts. Theoretically indebted to Harvey Molotch's growth machine thesis, the book essentially documents the protracted struggle between environmentalists and neighborhood activists on one side, and developers and pro-developer City Council members on the other. It argues that environmentalists beat the developers by electing an environmentally friendly City Council in 1997 and, in the process, made the environment the third rail in local politics.

Most of Austin's liberal elites were present at the party. I counted two present City Council members, several former council members, and many people employed at City Hall, Austin Energy, and numerous nonprofit, progressive, primarily environmental groups. Most of the attendees were white, appeared to be middle-aged professionals, were wearing T-shirts and flip-flops or casual business attire, and were drinking alcohol; it was happy hour, and local beers were only one dollar at the bar. Lots of people were eating fried chicken and burgers, although I would guess that many others were vegetarians or pescatarians, and the atmosphere was very convivial.

I could stay for only about an hour, enough time for a beer; unfortunately, I missed many speeches because I had to go across town to Austin's East Side for an event held by the environmental group People Organized in Defense of Earth and Her Resources (PODER). Founded at almost the same time as SOS, PODER is an environmental justice group, led mostly by women (many of whom identify as mothers), that primarily handles environmental issues facing Austin's East Side communities of color. PODER was not a part of the coalition that helped form SOS. Despite its small size, the organization is politically significant locally and recognized nationally as a noteworthy environmental justice organization; the Ford Foundation gave it a grant for its innovative work, and a nationally syndicated PBS documentary series profiled its successes. On that night, they were having their annual César E. Chávez "Sí Se Puede!" Awards Dinner at the Conley-Guerrero Senior Activity Center to recognize people who had helped them, and I was receiving an award for the work my students had done as part of a service-learning component of a geography capstone class.

Needless to say, the second party was very different from the first. While it was held in an austere community center, most people were dressed up, and the crowd was more diverse in terms of color and age. Most people were Hispanic, but there were also a significant number of African Americans and white people. I knew many attendees, but nobody was present from the City Council or Austin's white political and social establishment, although one woman there to receive an award had been at Swearingen's book signing. The people being honored were, for the most part, not college professors or urban professionals but people from the community that PODER serves in East Austin. They were clerical workers, teachers, and community activists. Many were teenagers and college students who had done work with PODER. For dinner, we had a choice of vegetarian or nonvegetarian tamales and beans, as well as chips, two kinds of salsa, and iced tea or water; there was no alcohol. Everything was homemade, and PODER's staff and volunteers had decorated all the tables.

The two events had different purposes; the former was a celebration of the accomplishment of an important figure in Austin's environmental community, and the latter was a dinner of appreciation for those who had helped an environmental organization. I think it is significant that the people I spoke to at the PODER dinner and Swearingen's book signing were largely unaware that the other event was happening. Perhaps each group had just failed to inform the other, but I would note that PODER is not mentioned once in Swearingen's history of Austin's environmental movement (an omission that, Swearingen claims, was due to "a matter of space") (Ankrum 2010a). Thus it would seem possible to tell the story of Austin's environmental movement without reference to PODER but impossible not to mention sos; in the same way it is possible to have an sos party for the environmental community without a large presence from PODER. I believe this contrast suggests that for a number of reasons, PODER and its constituencies remain largely outside the vision of the city's official environmental establishment, which, in Austin, means outside the inner circle of the city's liberal elites (Yznaga 2010). Not all people are blind to this omission, and I do not want to overstate the conflict; the two constituencies have often worked together (for instance, during the Tank Farm removal and sos campaigns, which I will discuss later). Yet the astigmatism, I would argue, largely remains intact, and as a result, what counts as an environmental concern, *a fortiori* as a liberal concern, is decided by certain kinds of people—namely educated white people who live in western Austin—and some "environmental" concerns, especially criticisms of the contemporary redevelopment of downtown, might be absent from the concerns of or even unintelligible to Austin's liberals. The lacunae in Swearingen's book and the groups' unawareness of the other's celebrations show the deep disconnect that exists between these two environmentalisms. Although their joint concern for the preservation of the "natural environment" has often unified the two groups around particular issues—overriding the cultural, class, and ethnic divisions that could have pulled them apart—the more intractable issue is the tension that arises from their different analyses of what counts as an environmental problem.

Urban Sprawl, Austin's Environmental Community, and Smart Growth Initiatives

While Austin had grown at a steady pace since the 1950s, since the 1980s its rate of growth has been astonishing. The region's population has more than doubled since the 1990s and is projected to grow even faster in the next twenty years (Steiner 2008). As in other Sunbelt cities, the steady rise in population is due mainly to immigration rather than natural increases, and it is driven by the development of the region's economy. Moreover, the total urbanized area in the region has also expanded at a phenomenal pace over the last thirty years, in part because many newly arriving residents have settled in recently built-out suburban areas or in burgeoning satellite cities that surround Austin's older neighborhoods. As in previous years, these spatial patterns of growth have been largely influenced by three interrelated factors: the prospecting of private developers in the nonurbanized areas; the support of some City Council members; and the state government's construction of highways.

As chapter 4 discussed in detail, a strong environmental movement has resisted the sprawling patterns of urban growth in Austin. In the mid-1980s, Austin's City Council had become clearly divided between two dominant political factions, suburban developers and environmentalists, and there was very little agreement about the future direction of city's growth. Despite efforts by the business community, Austin's environmental movement continued to grow in strength and even had some substantial victories. For example, in 1991 environmentalists ran a successful referendum campaign for the adoption of a local water-quality ordinance; the effort received strong support across Austin and was passed by the electorate by a decisive margin (Swearingen 2010).

By the late 1990s, the two factions were beginning to arrive at a compromise that came into being after the election of the "Green Council" in 1997, a moment that is seen by environmentalists as the high-water mark of their political influence. Significantly, it is also the critical turning point in the formation of a new relationship between environmentalists and the business community—a relationship that was embodied in the leadership of the then newly elected mayor, Kirk Watson. Watson offered a third-way platform of "Smart Growth" based on the three pillars of economic prosperity, ecological preservation, and social equality that echoed President Clinton's Council on Sustainable Development, which stated: "Our vision is of a life-sustaining Earth. We are committed to the achievement of a dignified, peaceful, and equitable existence. A sustainable United States will have a growing economy that provides equitable opportunities for satisfying livelihoods and a safe, healthy, high quality of life for current and future generations. Our nation will protect its environment, its natural resource base, and the functions and viability of natural systems on which all life depends" (President's Council on Sustainable Development 1996). Armed with the language of Smart Growth, Watson, known to be an especially deft negotiator, began to forge a way forward and brought the two competing factions to the table.

The essence of the compromise was to refashion an abandoned comprehensive master plan from the 1980s that called for steering growth off the Edwards Plateau and closer to the city's central business district. Bill Bunch, a leader of SOS, noted:

> Keep this area [pointing on a map at the Hill Country] as low density and protected as we can by steering our more intensive urban development to the east and downstream of the Edwards aquifer and along [what the 1979 comprehensive plan called] the preferred growth corridor. [This recommendation from that plan] then was sort of restyled during the green Watson council as the desired development zone. . . . Build here and preserve here [pointing to the western and eastern areas of the map]. This is your water supply to support your cities. . . . Preserve this, water and unbelievable biodiversity that still existed, species that live here and nowhere else in the world. Build here, this was the Blackland prairie mostly, it had already been denuded of its biodiversity by the plow and it is suitable for building. (Bunch 2008)

Following a similar vision, under Watson the City Council released a map that divided the city into three different development zones: a drinking water zone, an urban desired development zone, and a desired development zone

(Clark-Madison 1998). The idea was that development in most of west Austin, particularly the areas over the Edwards Aquifer, should be limited because of its potential impact on water, and that the City Council should promote infill and higher densities in the desired development zones, particularly the urban desired development zone, which covered virtually the entire eastern section of the city close to the central business district.

These designations were not codified into city's zoning regulations and therefore not enforceable as a matter of law, so instead City Council had to give incentives to developers to support the plan's goals (McCann 2003, 168). In 1997, three bonds that contained three elements of Austin's Smart Growth plan were proposed in one package, and their passage required the support of a diverse group of stakeholders. The first bond supported social equity and would allocate $10 million for remediating flooding risk along Walnut Creek in the Crystalbrook neighborhood (a more modestly resourced community in East Austin that was primarily home to African Americans). The second bond was designed to generate economic growth. It would raise the city's hotel room tax to 15 percent to finance an enlargement of the Austin Convention Center. In addition, these funds would be used to build a flood-diversion tunnel along Waller Creek that would remove a large portion of valuable land in the central business district from the floodplain and make it available for redevelopment. The final bond, intended to appease environmentalists, authorized raising $65 million from a water utility surcharge to buy 15,000 acres of undeveloped land in Austin's western suburbs. All the bonds passed, in part because of the strong support they received from the principal institutions and persons in both the business and environmental communities (Lindell 1998).

Austin's East Side and PODER's Challenge to Smart Growth

The redevelopment of Austin's central business district was largely interpreted as positive for both the environmental and business communities. "In the late 1990s, we were sort of more cooperating," Bill Bunch recalled, "where downtown was wanting to sell itself more and we were like, 'Yeah absolutely.' We are all for that" (Bunch 2008). The main vehicle for the promotion of downtown was a private-public partnership mainly consisting of a coalition of downtown land and business owners known as the Downtown Austin Alliance (DAA), which was financed by a special TIF (tax increment financing, using forecast tax gains to subsidize present-day improvements) on downtown real estate. The organization was designed to be "downtown's advocate," and the idea for creating it came from an investigation undertaken by the city, lobbied by the Chamber of Commerce, which had assessed the potential for redeveloping downtown to help it "regain its position as the economic hub of the city." The report noted,

"Downtown, however, is no longer perceived solely as a center of commerce but must also incorporate culture, entertainment, tourism and residential neighborhoods to maximize its chances for prosperity" (City of Austin 1992).

The changes that were being discussed by the DAA and the local government were significantly transforming the role of the city's downtown, which, for much of the twentieth century, had been a buffer zone between Austin's east and west communities. Austin had become more and more spatially segregated by race during the twentieth century (see chapter 6), with the majority of African Americans and Hispanics concentrated in neighborhoods just east of the city's central business district and wealthier whites to the west and north, a result of a combination of private and public forces. Located along the north bank of the Colorado River, east Austin contained some of the oldest neighborhoods in the city. As newer neighborhoods were built in other parts of the city during the 1920s and 1930s, price was an important impediment to residential mobility, but racially restrictive covenants in these newly developed areas also limited housing opportunities for nonwhites. At the same time, the municipal corporation imposed industrial and residential zoning standards and financed infrastructural improvements such as the development of a sewage system, water lines, parks, and schools that substantially altered the city's morphology. These measures also had a dramatic effect on racial and class settlement patterns, in part because much of east Austin's residential communities were zoned "miscellaneous or colored" and not "white residential," which subsequently left these communities open to industrial and commercial development that had been restricted in other, whiter parts of the city (City of Austin 1955). Additionally, when the local government made infrastructural improvements intended to benefit nonwhite minorities, such as building a school or hooking up electricity to a private residence, it would do so only for minorities living in east Austin and not for nonwhite communities in other parts of the city.

While for much of the 1980s environmentalism had developed primarily as a "west Austin" concern, in 1991, around the time SOS was formed, a group of people formed People Organized in Defense of the Earth and Her Resources in response to very different kinds of environmental stresses (Wallace 1994). Susana Almanza recalled how PODER was not always considered a part of the environmental community:

I can tell you when we first started working on the issues in our community and the environmental people would actually say they did not think that [the construction of high technology facilities in east Austin] was an environmental issue because basically at that time it was only nature kind that they were looking at—the water, the air, the animals—so people did not really figure into the environment. Their definition was nature kind and our definition was nature kind and humankind that you could not separate. . . . So we were saying, "No these are environmental issues because you impacted a community by building these facilities and you are using up this money [as a subsidy to spur development]. That is an environmental issue and you are impacting these people." The endangered part, we said, instead of the yellow cheeked warbler, we have people of color who are endangered. So if you want to look at endangered species we have people who are very endangered because of where we live and proximity to [the industrial development]. It took some education to get people to understand that these issues were environmental issues and to expand their scope of the environment to environmental justice issues. (Almanza 2010)

PODER's efforts began with a focus on the potential environmental and community effects of the rapid industrial development in the early 1990s of a corridor in East Austin, which contained a number of factories of high technology firms. Initially, the group was agitated about the local government's use of tax incentives to bring businesses that had minimal positive impact on the surrounding community in terms of employment. Moreover, toxic waste and heavy metal contamination were particular concerns. Chemical leaks and industrial accidents in 1982 and 1984 at Austin's Motorola plant, located in this corridor, had exposed these potential risks (Mora 1983; Cox and Banta 1984). Yet as PODER and its allies noted, nearly all the manufacturing plants of high-technology firms (and their associated risks) were located in East Austin around high concentrations of nonwhite minorities, while their research and management facilities were located elsewhere in the city (Hartenberger, Tufekci, and Davis 2012, 76–77). Furthermore, the potential long-term risks from toxic pollutants to these surrounding communities had never been explored, despite the fact that public EPA documents showed that more than two-thirds of the toxic pollution in these areas of Austin came from Motorola's and AMD's facilities (Citizen Fund 1990). Animated by these issues, in the 1990s PODER led a coalition of groups in a campaign that successfully forced Sematech to halt the use of some toxic chemicals in its East Austin production facility because of their potential health risks to the surrounding communities (Smith 1992; Mitchell 1993; Byster and Smith 2006).

PODER's local and national notoriety came later as the result of a successful campaign to have a tank farm, owned and operated by six major oil companies, relocated from East Austin (Scheibal 2002). In 1991 Mobil proposed expanding their facility, and PODER, along with community allies in both East and West Austin, organized to block the expansion. Local residents had long complained about the health and environmental impacts of these oil facilities—asthma rates in the surrounding area were very high—and testing revealed that the soil and water around the tank farm were highly polluted. After a long and protracted fight, the community finally signed agreements with all of the companies to move, and an environmental remediation fund and plan were established to clean up the site.

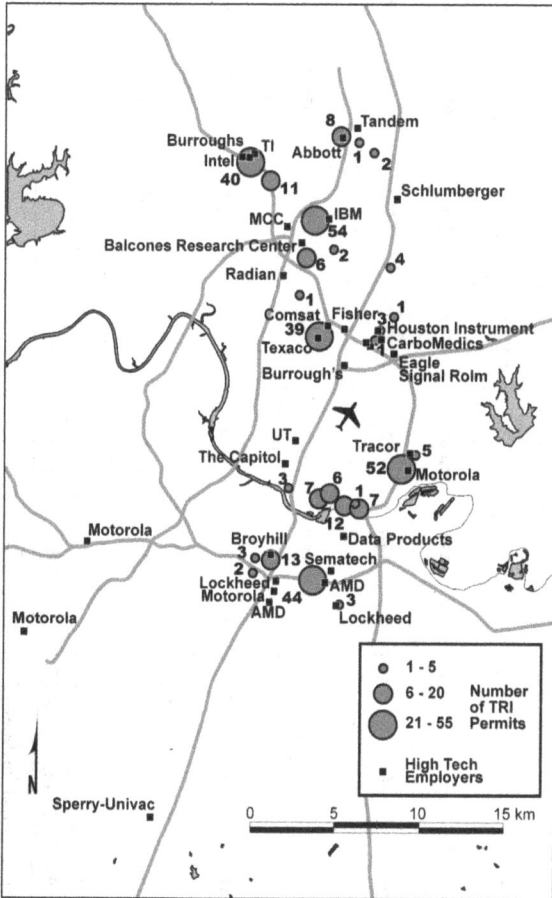

FIGURE 10.
The Number of Toxic
Release Inventory
Permits and the
Location of High-
Technology Employers
in Austin, 1987–1990.
Based on map of Toxic
Release Inventory Data
and High-Technology
Firms provided by
M. Anwar Sounny-
Slitine.

PODER's ability to mobilize people in their community during the 1990s mirrored the growing significance of the idea of environmental justice at a national
level. "Environmental justice" as a movement had gained widespread recognition during the 1980s, especially after the 1982 protests over a PCB landfill in
Warren, North Carolina, turned violent and projected the movement onto the
national stage (Chiro 1998, 108–113). One significant outcome of these protests
was a federally funded study that showed a spatial correlation between toxic
waste sites and racial and economic status through the southern United States.
A more robust social movement really galvanized, however, in the 1990s, after
the 1991 People of Color Environmental Leadership Summit, when hundreds
of organizations met and adopted the "principles of environmental justice." Often touted as "the most important single event in the movement's history," the
summit provided a forum to connect many organizations that had been working on similar issues in their local communities (Bullard and Johnson 2000,

TABLE 1. Central East Austin decadal demographic data by race and census tract, 1980–2010

Neighborhood (Census tract)	1980			1990			2000			2010		
	Non-Hispanic White	Non-Hispanic Black	Hispanic	Non-Hispanic White	Non-Hispanic Black	Hispanic	Non-Hispanic White	Non-Hispanic Black	Hispanic	Non-Hispanic White	Non-Hispanic Black	Hispanic
S. Upper Boggy Creek (4.02)	676	1,995	42	648	1,255	532	1,005	1,103	757	1,337	599	552
N. Govalle (8.01)	201	263	1,349	88	282	1,330	119	221	1,494	249	171	1,062
Rosewood (8.02)	27	3,135	477	108	1,960	659	116	1,798	1,328	535	1,200	1,283
Chestnut (8.03)	168	2,366	273	155	1,471	369	250	1,093	971	1,037	611	691
N. Central East Austin (8.04)	84	2,339	412	69	1,546	876	142	1,107	1,193	577	670	993
S. Central East Austin (9.01)	41	1,060	1,264	67	948*	868	152	622	1,013	628	475	862
N. Cesar Chavez (9.02)	304	398	5,273	265	193	4,694	218	404	4,673	1,008	409	3,902
S. Cesar Chavez (10)	784	56	4,527	559	59	3,929	630	87	3,872	1,092	87	2,395
TOTAL	2,285	11,612	13,995	1,959	7,714	13,257	2,632	6,435	15,301	6,463	4,222	11,740

Source: Data for census tracts from Neighborhood change database, using 2010 tract boundaries.

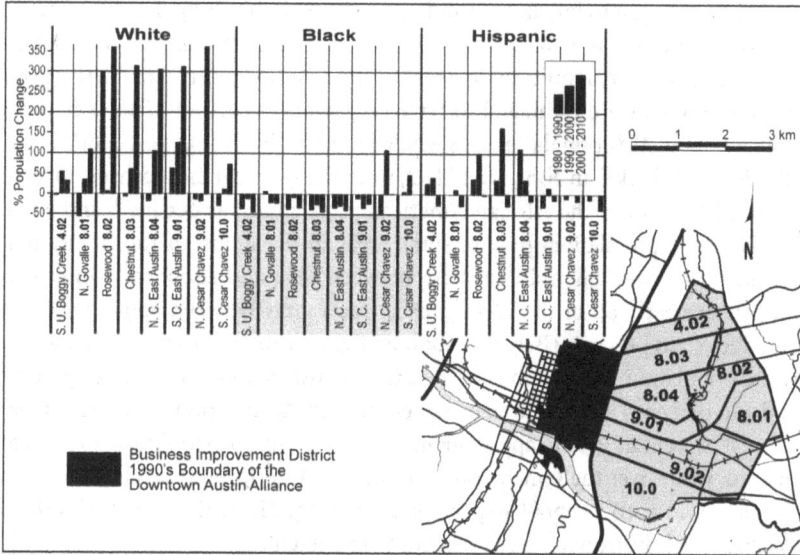

FIGURE 11. Central East Austin Decadal Rate of Demographic Change by Race and Census Tract, 1980–2010.

556). Additionally, federal and state governments enacted laws, the most famous being President Clinton's "Environmental Justice" executive order 12898, which required every federal agency to "make achieving environmental justice a part of its mission" (Clinton 1994; Cutter 1995, 113–114; Rast 2006, 253–254).

Responding to this federal concern and local pressure, in 1997 the city of Austin conducted a study to examine housing and industrial uses in East Austin. The results showed that East Austin had the greatest proportion of land zoned industrial in the city and that many houses were located on land zoned for commercial and industrial uses and near industrial facilities. The findings revealed the underlying environmental racism that resulted from relying on city zoning rules and plans that were developed in 1931, when it was a widely accepted idea to "concentrate what white planners felt were the unpleasant parts of the community—people of color and industry" (Greenberger 1997a). The report did not mention the long-term health effects of having generations of people living so close to human-made environmental hazards such as chemical effluent or industrial plant emissions, a point that PODER had been making, but the study was used by PODER and other groups to successfully lobby the City Council to change the zoning in the area to be more compatible with residential uses in future developments.

Other private and public forces were also working to change land-use patterns in East Austin. Private housing developers were quick to capitalize on East Austin's relatively inexpensive land prices and proximity to the city's central business district, and more and more relatively wealthy people (many of whom were white) began to buy houses and renovate them, resulting in the very rapid gentrification of some areas. This trend was particularly stark in central East Austin and was in significant contrast to the 1980s, when central East Austin neighborhoods were estimated to have lost about 18 percent of their population, while Austin's population as a whole increased by over 35 percent (Wilson, Rhodes, and Glickman 2007, 11). In fact, since the 1980s, in most census tracts in central East Austin, the African American population has declined, and while the Hispanic population grew in some areas in the 1990s, in the twenty-first century it too has fallen. In contrast, between 1990 and 2000, the white population in nearly all of the area's census tracts grew between about 7 percent and about 130 percent. (Between 2000 and 2010, this trend was starker, as the white population in the same areas grew by between approximately 30 percent and 360 percent.) The white share of central East Austin's population rose from about 8 percent in 1990 to approximately 11 percent in 2000 to almost 30 percent in 2010, while over the same period the African American share fell from about 34 percent in 1990 to about 19 percent in 2010; the Hispanic population's share declined slightly, from about 57 percent to 52 percent.

The city's Smart Growth initiatives indirectly supported this demographic shift by encouraging these new residential developments: almost the entirety of East Austin close to downtown was in the desired development zones. Drawing on the work of other environmental justice groups, PODER argued that the city was being reorganized using environmental principles that excluded communities of color (Almanza, Herrera, and Almanza 2003). Susana Almanza said:

> We didn't know what was happening [regarding the Watson's Council's plans to create Desirable Development Zones]. . . . As a matter of fact we stepped into it by accident because we were supposed to be at City Hall for a meeting and we kind of went into the wrong room and saw a big map about the Desirable Development Zones. That map showed moving everything east in Austin. We were like, "What's going on with this?" We were not at the table at the beginning. . . . We did not find out until it was just about a done deal. We went to City Hall to talk about it and say we were not in agreement with the way this was happening because it was moving industry further into East Austin. . . . We were very much against the whole Smart Growth movement because the Smart Growth movement was, to us, really the change of language from when they came in with the revitalization. . . . When they revitalized, that meant getting rid of us, making new communities, and we were not going to be in those communities. (Almanza 2010)

The frustration among long-term residents of the area toward Smart Growth was intense, and no Hispanic organization on the East Side came out in support of the Smart Growth bond initiatives; one group, El Concilio, vociferously opposed it (Leon 2010; Wear 1998). While Austin's environmentalists considered the Tank Farm and the concentration of industrial zoning in East Austin environmental issues, they did not believe the same about gentrification and the larger embourgeoisement of the area. Comments like the one made by longtime activist Paul Hernandez that the "SMART" in Smart Growth stood for "Send Minorities Across the River Today" meant nothing to environmentalists because they considered the transformation of East Austin's demographics and physical landscape to be social, not environmental, problems and part of the "natural" evolution of the city (Gandara 2002).

PODER's criticism of the Smart Growth plan was partly justified because while the plan offered a comprehensive vision for future growth, it largely ignored how asymmetrical power relations between whites and nonwhites had shaped the city's past growth. Part of the privilege of being white in Austin's past had been the ability to choose where you lived. Minorities, meanwhile, had nearly always been acted upon by political and economic forces and were routinely told (without much say in the matter) that their neighborhoods needed to be improved. For this reason, Hispanic activists, despite the potential upside and novelty of the Smart Growth plan, were able to criticize it as just another iteration in a longer historical process, like the urban renewal programs discussed in chapter 2, where communities of color had been pushed around in the city. To the extent that the plan offered a utopian sustainable vision for the future, it did so only by ignoring the dystopian history of racism and underdevelopment that, ironically, had resulted in East Austin's becoming a potential Desirable Development Zone.

In 2003, the Austin City Council officially abandoned its Smart Growth platform, in part because of the financial failure of several subsidized development projects in the central business district (Osborne and Scheibal 2003; Barna 2002). However, it was replaced with a new set of rules that looked very similar. The City Council still tried to provide incentives for downtown revitalization, but the plans were slightly modified to incorporate the principles of New Urbanism and its concern for affordability (Scheibal 2005). Needless to say, the basic outlines of the Smart Growth initiatives continued, and they continue to affect development throughout Austin and particularly East Austin, where more and more private housing and mixed-used developments have sprouted up, leading to a wholesale transformation of the physical environment and fundamentally altering the demographics of the area. Despite the rapidness of these changes, Austin's East Side continues to be distinguished by strong and vibrant communities of color, but how long these communities will survive becomes less clear every day. What is clear is that the city's planners are wagering that the environmental and economic benefits resulting from the redevelopment of the area—nearly every new luxury housing project touts the ecological benefits of being near Austin's downtown—are much greater than any potential social or economic value the longstanding forms of cultural difference could offer to the city's long-term competitiveness.

Conclusion

PODER's ability to attract the support of Austin's liberal elites had a lot to do with the organization's tactical use of the term "environment." In Austin, I have argued, environmentalism is the dominant ideology held by the liberal establishment, which for the last forty years has been the main oppositional force to the business community's control in local governance. Environmental justice organizations throughout the United States have been very effective at showing how environmental racism has influenced urban development by building on the power of pre-existing civil rights coalitions that were concerned primarily with racial equity. In contrast, PODER's success at addressing the problems of human-produced hazards, the results of past industrial zoning and racism, was due mainly to their ability to translate the issue from a social problem into an environmental problem, subsequently enabling them to win wide support across the environmental, and broader liberal, community. However, the group and its supporters have been largely unable to replicate this victory in their criticisms of gentrification and transformational developments to the east of the central business district because they have failed to make them intelligible environmental issues. So while this study points toward an important conjunction between environmental sensibilities and gentrification, or free-market solutions to neighborhood revitalization and upgrading, it also suggests that part of the problem (and perhaps the solution) concerns the manner in which environmentalists circumscribe what types of problems are environmental problems.

More than a decade ago, David Harvey lamented the inability and unwillingness of many environmentalists to effectively understand urban processes and incorporate social problems in their criticisms of urbanization; he called this failure "pernicious." But he still noted that the "integration of the urbanization question into the environmental-ecological question is a *sine qua non* for the twenty-first century" (Harvey 1996, 429). Unfortunately, the literature and policy formulations on urban sustainability and Smart Growth largely replicate the impasse found in Austin: urban and social concerns cannot be rendered as ecological and environmental problems. Figuring out forms of translation will help solve this problem, although it is not just an issue of translation: sometimes it is easy to avoid learning a new language when you do not need to learn it in order to survive, or when not knowing it makes it easier to not hear who is speaking. To confront the range of urban problems, we need to learn how to speak and listen effectively to the variety of ways in which the environment is described and understood. In this instance, it would require understanding how asymmetrical power relations, in both the present and the past, continually influence how and what kinds of "environmental" issues are addressed. Perhaps learning new ways of translating would help bring about the possibility for the creation of a more attentive "ecology of urbanization," one born from the melding of the environmental justice movement with mainstream environmentalism.

Yet if Harvey is correct, then thinking about the ways these two groups of environmentalists have worked together in Austin suggests it is possible to find integrated solutions. Many of Austin's liberal elites, including members of the City Council, are supportive of PODER's demands and are aware of the racial, class, and ethnic tensions that surround the push toward creating a more environmentally sensitive form of urbanization. Many predominantly white groups from Austin's West Side do take into account the concerns of PODER and other minority organizations, and while friction remains, these organizations argue that preserving ethnically diverse, affordable, and livable communities is necessary for engineering a more sustainable city.

Finding ways to achieve a win-win-win balancing of economic, environmental, and social concerns has been presented as the main problem in the struggle to create more sustainable cites. However, the main debates in Austin took place primarily between those who had an economic-environmental interpretation of urban sustainability and those who had an environmental-social one. The tensions between these two visions have become particularly stark because the business community has effectively appropriated a vision of environmentalism that does not treat the transformation of the urban environment as an environmental problem. Environmentalists need to challenge this selective definition of environmentalism, which permits a bifurcation of the movement, and instead offer a more expansive, more comprehensive definition that includes the urban environment. As long as social and ecological concerns are treated as distinct problems that have distinct solutions, rather than coconstitutive of a socioecological urbanization process, it will be possible to understand the transformation of the urban environment as something other than an environmental matter, and some environmentalists will continue to believe that a healthy approach to the environment is possible without a firm commitment to social justice.

CHAPTER 6

What Is Past Is Prologue

Urban Governance, Comprehensive Planning, and Political Reform

> The horizon open to the future, which is determined by expectations in the present, guides our access to the past. Inasmuch as we appropriate past experiences with an orientation to the future, the authentic present is preserved as the locus of continuing tradition and of innovation at once; the one is not possible without the other, and both merge into the objectivity proper to a context of effective history.
>
> Jürgen Habermas, *The Philosophical Discourse of Modernity: Twelve Lectures*

> As Jackson's second term came to a close, political arrangements in the city remained somewhat unsettled. Jackson had been unable to convert his electoral base into a progressive governing coalition. The business elite had too many strategic alliances to be pushed aside.
>
> Clarence Stone, *Regime Politics: Governing Atlanta, 1946–1988*

For most of the twentieth century and beyond, major planning and political reform efforts in Austin have been interdependent. The major reorganizations of the local government in 1926 and 1951 were accompanied by new master planning endeavors, while the failure of the city's comprehensive planning efforts from 1977 to 2011 were concurrent with a series of unsuccessful attempts to alter the at-large system of representation on Austin's City Council. In 2012, this tendency for confluence appeared again: Austin's City Council adopted a new comprehensive plan in June, and in November the city's electorate overwhelmingly supported changing from an at-large council-mayor-manager system to a single-district council-mayor-manager system.

One crucial reason for the success of both of these recent initiatives was the explicit or tacit support (or nonopposition) of key organizations from the business community, and I would argue that this reveals an enduring quality about urban governance in Austin: when the business community's leading organizations (chiefly the Chamber of Commerce) support a major reform in the city it is more viable, and when they do not it tends to languish. While this is not a novel observation, I think it suggests that the business community's principal representatives still have much more sway in Austin's governance than other contemporary accounts might indicate (Long 2010; Moore 2007; Straubhaar et al. 2012; Swearingen 2010). Does my contention imply that none of City Hall's agenda is achieved without the support of the city's leading business groups? No. But I think that the history of urban governance in Austin indicates that it can be much harder to implement broad reforms or programs without the support of the business community's most important associations.

Of course, this problem has preoccupied many scholars concerned with the political economy of cities: how, in democratic societies, can we account for the asymmetrical influence the business community's principal associations have in urban governance, especially when the priorities of a ruling urban coalition do not correspond with the preferences of the business community's chief organs? One quite original answer comes from regime theorists, particularly Clarence Stone. For Stone the ability of an urban coalition to formulate and implement its agenda—to govern—results from the strength of the "informal arrangements by which public bodies and private interests function together in order to be able to make and carry out governing decisions" (Stone 1989, 6). Importantly, private interests, in this context, include groups that represent any politically interested party—such as labor unions, environmentalists, historic preservationists, and ethnic groups—and not just business interests. Nevertheless, the business community's leading organizations and their development priorities often become the central concern of city governments. This happens not because a cabal of business elites coerces City Hall but because they control a much larger stock of resources that can be brought to the table and used, in conjunction with the local government, to forge cooperative relationships across institutional boundaries, which might otherwise impede working among groups, toward a concerted end.

Although an urban coalition may be elected that controls the local government's formal machinery, it may fail to deliver on its priorities—fail to govern effectively—because the coalition has been unable to mobilize and coordinate available resources for a concerted end and to overcome internal divisions and external opposition. Not being able to govern effectively does not mean that nothing happens; in Austin, for instance, as this chapter shows, many attempts were made to reform the city's governance and refashion its urban form, and many programs and policies were enacted that had significant and long-lasting impact on the city. It seems, though, that without sufficient power to coordinate informally and formally with many communities of interest, to engage and harness their resources toward joint goals, a governing coalition's capacity to realize its central priorities will be frustrated, limited, or undermined.

The problem explored in this chapter is how reforms to formal election systems for local government went hand in hand with comprehensive planning initiatives. A considerable body of literature has focused on the legal and po-

Year Proposed	Name of Plan	If Adopted	Electorate Support for Bonding Priorities
1928	Koch and Fowler	Yes	Yes
1955	Austin Plan	No	Yes
1961	Austin Development Plan	Yes	Yes
1979	Austin Tomorrow	Yes	No
1987	Austinplan	No	N/A
1997	Neighbourhood Planning	Yes	Yes
2012	Imagine Austin	Yes	Unknown

FIGURE 12. Master Planning and Bonding Priorities in Austin, 1928–2010.

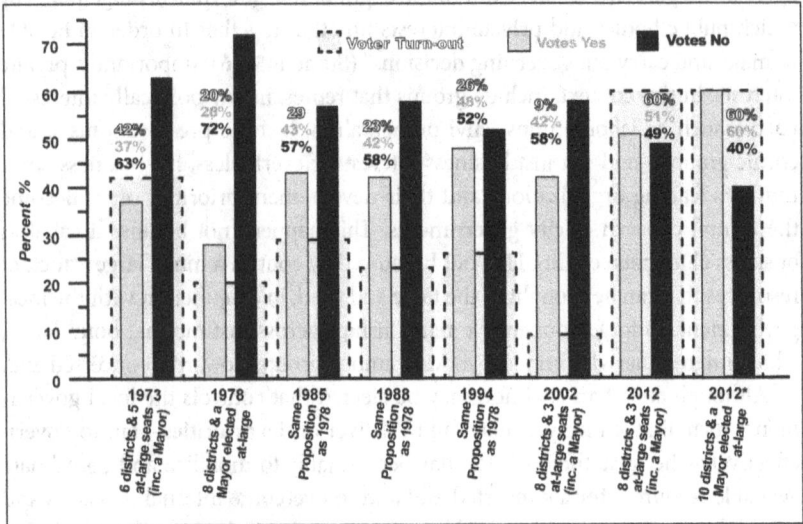

litical implications, especially for disenfranchising nonwhite communities, of the various election systems used by local authorities in Texas (Brischetto et al. 1994; Davidson 1992; Davidson and Korbel 1981). Moreover, the scholarship on comprehensive plans shows that planning is always a political act; it engages the formal institutions of government and reflects the ideals and interests of a large range of private interest groups (Hoch 2007; Ryan 2011). Yet there is surprisingly little discussion about how changing governing coalitions and institutional reforms to local government election systems affect the priorities set forth in comprehensive plans. This oversight is all the more unexpected given that there is a fairly developed literature about how official reforms in municipal representation at the start of the twentieth century affected the priorities set forth in comprehensive plans (Bridges 1997; Brownell 1975; Tindall 1967; Tretter 2013). Yet little has been made of how these two processes ran together at the twilight of the twentieth century and the dawn of the twenty-first, when there were both a push to unwind Progressive-era municipal reforms of election systems and significant attempts by cities to develop new comprehensive plans around the goals of urban sustainability.

In this chapter, I argue that an underlying system of power, which influences the structures of urban governance in Austin, can be revealed by exploring the different efforts at changing its system of municipal representation and in the development of its comprehensive plans. What this history shows is that the business community's principal organizations had an unusually strong role in influencing whether certain priorities were implemented by the city's government and adopted by the electorate. While coalitions not endorsed by the business community's leading associations have, at various times, controlled the local government's formal levers of power, without the support of the business community's chief institutions, they were not able to deliver on their major priorities.

Furthermore, I will show that by looking at the past we can see how significant the structure of Austin's system for electing local government officials is for controlling the city's urban agenda. In the 1920s and 1950s, City Council reforms were crucial to how comprehensive plans developed and set the priorities for urban development that were supported by bond packages. Moreover, the limited success of urban coalitions in implementing large parts of their reform agenda from the 1970s to the 1990s shows how interdependent the city's various electoral systems and governing capacity have been. In fact, I will go so far as to suggest that liberal reformers' failure during that time to maintain a governing coalition on the City Council and to enact comprehensive planning reforms was premised on their inability to abolish the at-large system. Had such an effort been successful, their ruling position would have been strengthened and one perpetual division that ultimately helped undermine the strength of this coalition, namely a racial or ethnic division, may have been overcome. Today, as Austin implements both significant municipal reform and comprehensive planning, it is worth looking at how both were accomplished because of the support of the business community's leading organizations and how institutional reforms may be reflected in the priorities that can be fulfilled under the direction of the new comprehensive plan.

Imagining Austin's Future:
Better Representation, Better Planning

On November 6, 2012, after nearly forty years of attempts, Austin's electorate overwhelmingly approved a ballot measure to abolish the city's at-large council member system and replace it with a system of single-member districts. The reform had been supported by the city's nonwhite minorities and neighborhood associations for decades, and in 2012 the primary push came from a diverse coalition that was strongly supported by the Hispanic community and neighborhood activists, from both central and suburban areas. These proponents argued that the City Council was largely made up of people who lived in selected neighborhoods, mainly those in Central West Austin, and that they therefore represented only the interests of those neighborhoods (Austinites for Geographical Representation 2012a).

FIGURE 14.
Percent of Ballots
Cast in Favor of 8–2–1
Proposition by Precinct.
Based on map provided
by Ryan Robertson, City
of Austin Demographer.

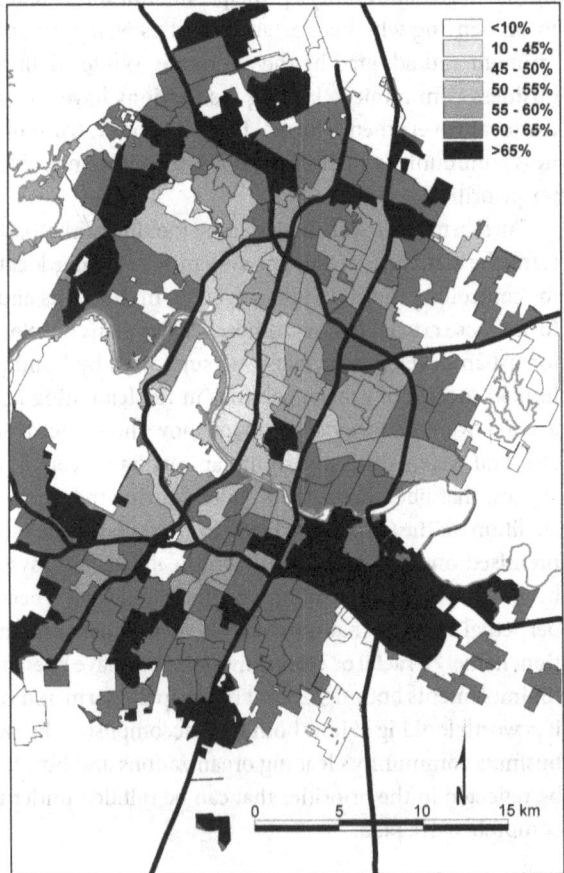

FIGURE 15.
Percent of Ballots
Cast in Favor of 10–1
Proposition by Precinct.
Based on map provided
by Ryan Robertson, City
of Austin Demographer.

One notable shift over the years was weakening opposition to this political reform from the business community's principal organizations. For nearly three decades the business community had formed an abiding opposition to this reform, but during the unsuccessful attempt to adopt a single-member district system in 2002 both the Chamber of Commerce and the Real Estate Council of Austin (RECA) changed their position (Clark-Madison 2002). In 2012, business groups took a much more aggressive stance in their support of this political reform and helped finance the campaigns that supported the ballot propositions. Granted, there were two proposals offered in 2012: "8–2–1" and "10–1." The first proposed creating eight single-member districts, two at-large council seats, and an at-large mayor and had much stronger support from the business community; the campaign that supported this proposal was primarily funded by RECA and a local property developer known as FM Properties (formerly Stratus) (Aus-

tin Community for Change 2012). The second proposal supported the formation of ten single-member districts and one at-large mayor, and it had a much
broader coalition of support, including significant financial backing from the
Texas Association of Builders and the Austin Home Builders Association (HBA)
(Martin 2012; Austinites for Geographic Representation 2013; Austinites for
Geographical Representation 2012b). While the Chamber of Commerce took no
formal position on either proposal, its chairperson did endorse the general idea
of some kind of electoral reform in principle (Fontenot 2012). In the end, 10–1
garnered more votes, so it is the system that was implemented, but if the two
proposals are taken together to be part of the same reform effort, then it shows
that the business community displayed strong support in 2012 for the proposed
reform.

In contrast, a belt of neighborhoods in Central West Austin, where in the
past there had been significant support for similar reform efforts, was the area of
its strongest opposition in 2012. It was telling that the West Austin Democrats,
one of this area's more important advocacy groups, was one of the few political
associations that publicly opposed both of the 2012 redistricting efforts (Sadun
2012). Certainly Central West Austin is wealthier and whiter than other areas
(although not the wealthiest or the whitest in Austin), and, as was mentioned
before, a disproportionately large number of City Council members and mayors
since the 1970s have come from this area. However, what might be as significant
is the relationship of these neighborhoods to Austin's mainstream environmental community. Over the last forty years, many people from these neighborhoods formed the core constituency in the mainstream environmental community, and this community was very concerned about the implications of this
political reform on their political power and, especially, their ability to influence
the future priorities of the City Council (Long 2014). This point was made, for
instance, by political consultant Marc Yznaga, who suggested that there would
be a better chance of forming a more probusiness (by which he meant antien-
vironmental) coalition on the City Council in a single-member district system
because numerous suburban and more conservative areas had been annexed
and incorporated into the political jurisdiction of the City of Austin over the
last three decades (Yznaga 2013). According to Yznaga, single-member districts,
while supporting minority constituencies within the city's limits, would also increase the clout of these suburban voters and dilute the power of the more progressive (more environmentally concerned) inner-city neighborhoods.

The same year Austin's electorate voted to change the city's charter, the City
Council adopted the Imagine Austin comprehensive plan. The final plan was
the result of a massive two-year "collaborative planning" effort guided by the
national design and planning firm Wallace, Roberts, and Todd. It involved
over eighteen thousand public participation inputs, which included a variety
of public meetings, events, and forums; several internet surveys; and hundreds
of do-it-yourself "meetings in a box." There were also significant contributions

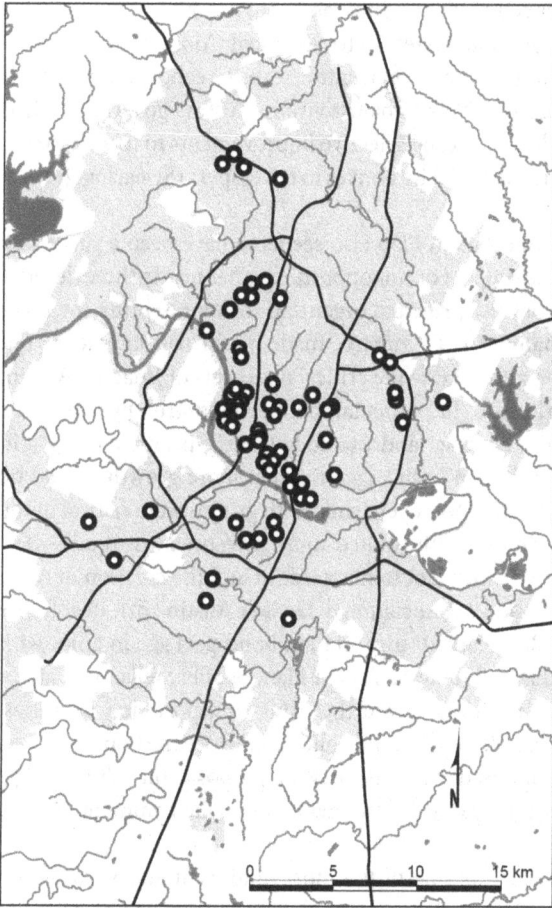

FIGURE 16.
Austin City Council
Members' Residences
1971–2010. Based on
map created by Ken
Martin at the Austin
Bulldog, 2011.

from private meetings with fifty-five stakeholder groups such as the Chamber
of Commerce, RECA, and the Austin Neighborhoods Council (ANC) (City of
Austin 2012). While the comments from the public inputs were significant, the
basic outlines of the planning effort and its vision were most strongly shaped by
the Citizen's Advisory Task Force and by two other formally appointed working groups, the Comprehensive Plan Committee of the Planning Commission
and the Comprehensive Planning and Transportation Subcommittee of the City
Council (City of Austin 2012).

To a large extent the 2012 plan reflected and codified the dominant New
Urbanist vision that had been pursued in the city's planning efforts over the
previous two decades (Barna 2002). These priorities were developed in various
initiatives and programs of different departments within the city's government
such as the Smart Growth Initiative (1998), the 2035 Sustainable City Initiative
(2006), and the Strategic Mobility Plan (2009) (Gregor 2010). The planning vi-

sion had also been influenced by the findings and plans of other regional organizations, most notably the Capital Area Metro Regional Coordination Plan for
the Capital Area (2006) and the Envision Central Texas Greenprint for Growth
(2009). Moreover, to fulfill this New Urbanist vision, Austin's government had
modified its land-use codes, providing selective improvements to its infrastructure, and had granted some financial incentives to developers (Swearingen 2010;
Barna 2002).

Nevertheless, the Imagine Austin Plan also seems to have been a successful
effort in part because of the support or nonopposition, after nearly three decades
of unyielding hostility, from the principal organizations representing the business community. The Chamber of Commerce made no formal statements, but
public comments by RECA and HBA, while critical of aspects of Imagine Austin,
generally endorsed its priorities. In particular, both groups strongly supported
the proposal to rewrite the existing land-use code to help promote density in
existing neighborhoods, although both suggested that more growth should be
planned for in Austin's suburbs. Moreover, while both groups were critical of the
plan's transportation priorities—both wanted more investment in roads—neither objected to a greater commitment to mass transit and the creation of more
mobility choices (Home Builders Association of Greater Austin 2011; Real Estate
Council of Austin 2011). But a large transportation bond package in 2010, with
priorities largely codified in the Imagine Austin Plan, possibly tells more about
the business community's position toward the plan's transportation priorities.
The Chamber of Commerce offered strong public support for the bond package,
and while RECA formally opposed the transportation proposal, they did not support the organized campaign against it; HBA made no public comment (Austin
Business Journal 2010).

It is also important to look at one of the most significant and vocal opponents to the 2012 Imagine Austin plan, a group that previously had been one of
the most ardent supporters of the city's comprehensive planning efforts: neighborhood association activists. Many neighborhood association leaders saw the
2012 plan as a reflection of the priorities of the business community. As a result,
the ANC, the umbrella organization for the many neighborhood associations in
Austin, refused to support it. Neighborhood activist Jeff Jack went so far as to
call it an "elitist plan," and even a member of the Imagine Austin Plan's Citizen
Advisory Taskforce, Ora Houston, called it a "land grab" (Toohey 2012; Smith
2012). The core issue these detractors had with the plan was that it envisioned
considerably higher densities and more growth in existing urbanized areas.
Many complained that politically connected areas, namely West Austin, would
be excluded from the plan's effects (although, as the next section shows, existing private zoning makes denser development in this area nearly impossible).
Because the plan anticipated more growth in central Austin neighborhoods, it
would lead to ever greater pressures on older neighborhoods, particularly rising

land tax burdens resulting from the increased exchange value of land that accompanies new investments and demographic growth. These activists believed that the final plan largely ignored their input, concerns, and suggestions and was basically, as former head of the ANC Cory Walton said, just a "huge bonus" for developers (Ankrum 2010b).

Both the reaction of the opponents and the available evidence about support show that the coalitions that backed both the political reform (to a system of electoral districts) and comprehensive planning efforts largely relied on the support of the business community's leading organizations, or at a minimum their nonopposition. While neighborhood activists that opposed the Imagine Austin comprehensive plan identified some problems with it, they were unable to stop it because of the coalition that formed between those organizations and some environmentalists. In the case of the political reform, those nervous that the plan might undermine the West Austin–focused environmental community's role in local governance were caught in the reverse situation: contending against these same business groups in a different coalition supported largely by outlying areas, neighborhood association activists, and the city's nonwhite minorities, particularly Hispanics.

Much of the discussion about the impact of the contemporary political and comprehensive planning efforts treats them as two separate and distinct elements of a changing city, failing to appreciate what the confluence of the two in the city's history exposes about the structures of governance in Austin. This history reveals the enduring asymmetrical influence of the business community in helping policies and reforms come to fruition. Focusing on how planning and political reform have converged during the twentieth century requires looking not only at the instances when reformers were able to implement their priorities but also when they failed.

Progressivism, Planning, and Jim Crow

In Austin, as in much of Texas, Progressive-era municipal reform efforts were generally led by the business community's chief organizations, which were opposed to ward politics because of its potential for "corruption" and its tendency to support more populist elements in government (Rice 1977). In Austin in 1908 "a group of citizens," led by the Business League, ran a campaign "to convince the [city's] voters that a change in the city charter providing for commission government would assist in finding solution to their many municipal difficulties" (MacCorkle 1973, 35). The following year, Austin adopted a new municipal charter, which changed the structure of its local government from a ward system to an at-large commission system (Staniszewski 1977, 55). In local elections the same year, a slate of reform candidates, backed by the leading members of the business community, won a landslide election and took control of the newly formed commission government (Austin Daily Statesman 1908; Staniszewski 1977, 27–29). Soon after entering office, the new administration undertook an effort to finance public works programs designed to encourage residential and commercial growth in Austin, such as building sewers and constructing a new dam for electricity (Stone, Price, and Stone 1939, 5; Seveik 1992).

Sixteen years later, in 1924, Austin adopted another Progressive-era reform and changed to a city-manager government; Austin was, in fact, the first larger city to do so. Unsatisfied with the 1909 reforms, members of the Chamber of Commerce, in an effort spearheaded by longtime manager Walter Long, in 1917 had begun advocating a change to a city-manager system (MacCorkle 1973, 48–49; Stone, Price, and Stone 1939, 16–17, 19). The principal reason given for this reform was the city commissioners' inability to secure a new bond package that would help Austin maintain or expand its municipal services (Stone, Price, and Stone 1939, 5–6, 14–15). While there was substantial opposition to changing the form of Austin's administration (the ballot proposition barely passed), in 1928, only four years after the change was finalized, the city's electorate endorsed a large bond package that helped finance capital improvements proposed in the city's first comprehensive plan (Austin Chamber of Commerce and Long 1948, 26). Much of the support for this bond package came from the two wealthiest areas of the city and the "Black ward," which was promised much-needed public investment in their community (Stone, Price, and Stone 1939, 20, 27; Staniszewski 1977, 33).

Like other "business progressives," members of Austin's business community believed that zoning was a means to modernize the city and make it a more suitable place for investment and commercial opportunities, and for about a decade, they actively tried to persuade the city government to embrace the principles of zoning and comprehensive planning (Long 1927; Long 1928). When Texas legalized comprehensive zoning in 1927 (largely modeled on the Standard Zoning Enabling Act), Austin was one of the first Texas cities to adopt a comprehensive zoning ordinance (Mixon 1991, 1:6–7; Chase 1932, 42). In addition, in 1928 Austin's City Council, at the behest of the Chamber of Commerce, hired the Dallas planning firm Koch and Fowler to create a master plan, which was adopted in 1929. The plan offered recommendations for public zoning regulations that encouraged separating residential, commercial, and industrial land uses. Under these new zoning rules, neighborhoods to the west, northwest, and north of the central business district were designated primarily for single-family residential developments, while those to the east and northeast were zoned for commercial, industrial, or unrestricted uses.

Antiblack racism played an important role in these early planning efforts. In Austin, as in many other cities in the South, ruling white elites wanted to create "racial zoning" standards—that is, to use racial categories in zoning and planning regulations—but court rulings routinely held such standards to be il-

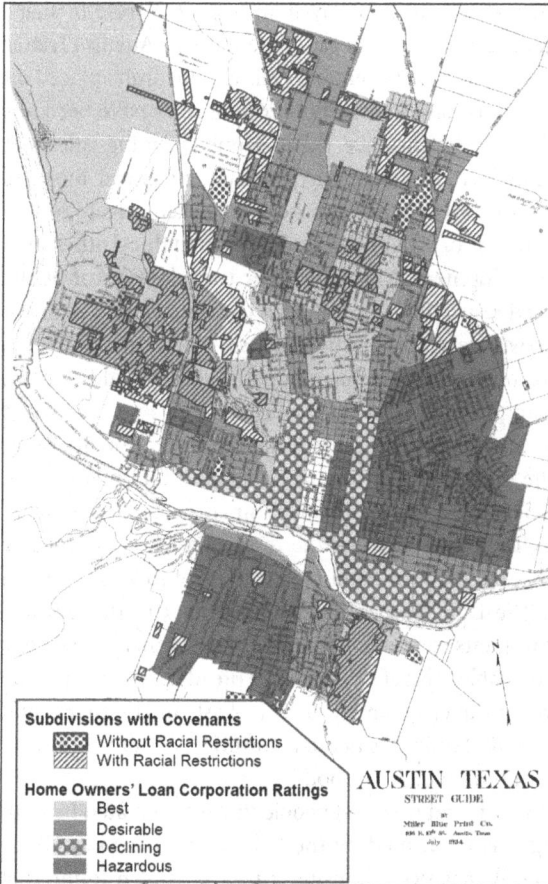

FIGURE 17.
The HOLC's Mortgage
Risk Map and Austin
Subdivisions with Racial
Covenants. Based on
map provided by M.
Anwar Sounny-Slitine.

legal, so most comprehensive plans began to rely on "race-based" zoning. In race-based zoning (using legally segregated public services to engender racial apartheid), city planners could encourage racial segregation by creating "racial districts" for nonwhites with comparable municipal services (Silver 1997). Austin's 1928 planning followed that course and suggested the creation of "a Negro district" in the eastern section of the city (City of Austin 1928, 57). So while bond money was used for capital improvements that helped African Americans, they were designed to benefit only those who resided in East Austin (Kraus 1973, 150–152). A similar, albeit not as extensive, approach was used to segregate Hispanics in East Austin despite the fact that neither their housing patterns nor their use of municipal services was mentioned in the comprehensive plan (McDonald 2012, 101–143). Importantly, the areas in East Austin designated for African Americans and Hispanics were the same areas that were not zoned exclusively for residential uses under the 1928 plan.

Private zoning by deed restrictions also played an important role in shaping long-term land uses and residential segregation patterns in Austin (Tretter 2013). In fact, since the beginning of the twentieth century, the limits placed on land use by restrictive covenants have guided how zoning rules have been applied to certain areas of the city. For instance, the 1928 public zoning rules that created commercial and industrial districts, particularly the Central Business District, were used only in those areas because those areas lacked private zoning restrictions on noxious land uses or strict rules prohibiting anything other than single-family residential development. Moreover, after the adoption of public regulations on land use, deed restrictions were applied to a much larger segment of the city, especially in newly developing subdivisions in the north and west, following the 1928 plan's recommendations for residential zoning. Additionally, nearly all residential housing deeds in Austin, especially after 1935, contained racial restrictions that prohibited African Americans (and occasionally other nonwhite groups) from the right to use or occupy land (except, in many cases, as domestic servants). When these restrictions combined with the system of segregated public services, the result was that the residential choices available to African Americans were extremely limited in many parts of Austin for much of the twentieth century. The U.S. Supreme Court ruled in 1949 that racially restrictive provisions in covenants were unenforceable, but all other restrictions on land use remained enforceable. Therefore, the widespread absence of private zoning rules in some parts of the city, namely areas in East Austin, made, and continues to make, non–single family residential, commercial, and industrial development much easier in those neighborhoods.

Moreover, the segregation of land uses and people in Austin created by both public and private zoning was reaffirmed by the federal government's Home Owner's Loan Corporation. A 1935 HOLC housing survey of Austin found that the city "suffered little" from the national economic decline, particularly in home values, and that "the greatest drop in values" was in "negro and inferior white sections in the southern part of the city" (Home Owners' Loan Corporation 1935, 8–9). However, the "Realty Area Map" showed all areas marked on the "Race Map" as the places where Hispanics and African Americans resided in large numbers, mostly in East Austin, in red, meaning they were "dangerous" for investment. The HOLC report was not a legally enforceable document, but it did "represent the composite judgment of seventeen well-informed real estate and mortgage men of Austin" (Home Owners' Loan Corporation 1935, 21). Their prejudice may have affected their willingness to lend money to homeowners in the area and, as a consequence, driven out financial opportunities, encouraged spatial segregation, and undermined the capacity of East Austinites to maintain the quality of their housing stock. Most importantly, the map reflected the influence of the 1928 Comprehensive Plan, as well as the pattern of distribution of private covenants. Marked in the 1928 plan as "white residential," western sec-

tions with covenants were later designated as the areas trending best for mortgage risk in the HOLC residential security map. It is worth noting too that the
section marked "best" on Austin's residential security map is also the section
where in 1950 the greatest percentage of racial and other forms of private land-
use restrictions could be found.

In Austin, in the first half of the twentieth century, successful attempts to
reform the city's system of elections became conjoined with its comprehensive
planning efforts. Leading figures from the business community had attempted
to remake Austin through comprehensive planning. They had modernized the
city's infrastructure, codified the preexisting residential and land-use patterns
through zoning controls, and solidified the racial apartheid within Austin. Most
importantly, these leaders understood that the formal system for electing officials to the local government was important to the realization of their vision for
the city. As the next section shows, this junction would continue through the
middle portion of the twentieth century.

Planning for the Industrialization of Austin

The at-large system put in place in 1909 to elect representatives to the City
Council (first called "commissioners" then later "council members") created five
seats; until 1971, these representatives then nominated a mayor. Out of any field
of potential candidates that ran for the City Council, the five individuals who
received the greatest number of votes were elected. While some city residents
worried that certain people or areas could have undue influence on electing particular City Council members, the principle behind at-large governance was to
make representatives accountable to a broader segment of the population. But
the effect of the system was to entrench incumbent candidates, make it more
expensive to run for City Council, and limit the influence of less wealthy and
territorially concentrated communities, such as nonwhite minorities, on local
government (Olson 1965, 12). In effect, it supported the status quo, perpetuating
the influence of the business community's principal organizations, which had
resources and an extensive network of representation across the entire city, and
of "white people," who were allowed to live in a larger number of zones throughout the city.

While the democratic "deficit" created by Progressive-era electoral reforms
had long been documented, beginning in the early 1950s there were increasing calls to change the general at-large system to an at-large place system—a
change that in fact reinforced the antidemocratic features of the at-large system
(Bridges 1997, 75; Martin 1933, 212). Advocates of this reform seemed to have
been motivated by the election of Emma Long to the City Council in 1948, although the nearly successful eighth-place finish of Arthur DeWitty, an African

American, in 1951 seems to have also provoked calls for this change. Long was elected in part because of her popularity with nonwhite minorities, and during her first few years on the City Council she "developed an active opposition in the community, especially among the bankers, land developers, and the large business people" (Orum 1987, 214). In 1952, business groups opposed to Long, plus a few others, began to press for an at-large place system. The idea was that five at-large places (or seats) would be created, and each candidate would have to choose to run for only one of these seats. The stated rationale for this proposal was to prevent the election of candidates who received less than 50 percent of the popular vote (which did happen) and to discourage "single-shot" voting, where voters cast votes for only one candidate rather than five (For A Better Austin 1953). Critics, however, charged that these reforms were designed to help "business-oriented candidates" and decrease "the possibility" that an African American would be elected (Olson 1965, 11). Despite these objections, made most vocally by Long, in 1953 the electorate, in a referendum, overwhelmingly approved this reform. Although Long maintained a position on the City Council (and served for many more years), the reform solidified the business community as a perdurable bloc in future elections.

Like the reforms of the 1920s, the reform of the City Council in 1953 occurred concurrently with attempts to increase the local government's involvement in urban planning. Along with the City Council reform charter amendment adopted in 1952, thirty-two other propositions supported by the business community won approval from Austin's electorate. Several of the propositions were designed to greatly empower the Austin Planning Commission. In particular, these reforms allowed the Planning Commission to recommend a master plan to the City Council for approval and made any adopted master plan a legally enforceable guide to the commission's future decisions. Pressure to create a master plan had been building from a coalition led by the Chamber of Commerce that included groups such as the League of Women Voters and the Austin Real Estate Board (Mitchell 1957). Under its new authority, the Planning Commission hired an urban planner to update the 1928 comprehensive plan, and in 1955 a detailed draft plan called the Austin Plan was released. It offered a completely new vision for the city, one that foresaw significant increases in Austin's population over the next three decades, large-scale industrial development, expansive residential and commercial suburbanization, the renewal of older sections of the city, and the changing character of businesses in the central business district (City of Austin 1955).

Many of the specific recommendations from the 1955 Austin Plan were later rejected, but the overall vision embodied in the plan was endorsed by the master plan adopted in 1961 and known as the Austin Development Plan. It envisioned the reorganization of the central business district to accommodate automobiles, large-scale suburban residential and commercial growth (especially in the north and northwest areas), an urban renewal program in the neighborhoods of more modestly resourced peoples, and extensive industrial development (City of Austin 1961, 5–10). Moreover, the 1961 plan proposed a vastly expanded network of arterial roads.

The 1961 plan also offered specific recommendations to support industrial growth. It noted that although Austin had some limited industrial development, it should be "attractive to new industries such as research, fabricating plants, electronics manufacturing, precision tool and instrument manufacture and other 'light' industrial operations" (City of Austin 1961, 9). To meet this future demand, the plan recommended that three areas in the city be designated as the places where expanding future industrial development should be concentrated. Two were in suburban locations, but the third was closer to Austin's downtown:

> The Land Use Plan anticipates the continued development of large industrial areas in the northwest, east, and southeast sections of the city. . . . It is anticipated that many industrial plants in the downtown area will find it desirable to move to outlying locations in the future. These industries, as well as present scattered industries, should be encouraged to relocate in planned industrial areas when possible. The present industrial operations occupy about 400 acres of land and related "heavy" commercial uses about 100 acres. The Plan envisions that about 3,400 acres will be needed to meet the requirements of industrial growth. (City of Austin 1961, 9)

To fulfill this industrial vision, the plan, clearly pointing to the land-use conditions just to the east of the central business district, called for the "redevelopment of industrial areas where scattered residences, small parcels, and inadequate streets inhibit industrial development" (City of Austin 1961, 10). In this respect, the plan's recommendations for where industrial growth ought to occur simply reinforced the patterns of land-use segregation that had been established in the 1920s by private and public zoning regulations.

Over the next decade Austin would undergo a dramatic expansion, and in order to enable the creation of the new industrial Austin, control over the implementation of the planning process was essential. Beginning in the 1950s, the Planning Commission, the City Council, and the electorate heavily favored Austin's rapid urbanization. Millions of dollars were spent on extending and improving existing public infrastructure to keep pace with the expanding city. To get a sense of this growth one need only look at the increase in Austin's municipal debt from 1945 to 1973. In 1945, the city's general obligation bonds were about $5 million (about the same as 1930), and by 1973 they were almost $45 million (about $18 million in 1945 dollars). Similarly, revenue bonds grew from about $50 million in 1960 to over $155 million in 1973 (about $100 million in 1960 dollars) (City of Austin 1974, 26). The significant increase in the city's bonding in the 1960s largely supported the priorities set down by Austin's 1961 master plan and its program of residential growth, urban renewal, and light industrial development (City of Austin 1965, 3; McMurtry 1963). By the early 1970s, all of these efforts had come to fruition. Ever greater portions of the urban fringe had been converted into residential housing, a publically funded urban renewal program was renovating many neighborhoods (especially in East Austin), and industrial development was finally being realized, as Chamber of Commerce efforts had proved effective at supporting homegrown industrial firms and bringing a branch of IBM to Austin in 1966.

As in the 1920s, establishing and realizing the priorities laid out in the city's 1961 comprehensive plan was connected to the implementation of the 1953 political reform because this reform had helped ensure that certain kinds of people and interests had more representation within the local government. Had nonwhites and more modestly resourced communities been more represented on the Planning Commission or the City Council, the master plan may have expressed a different set of priorities. Instead, city planning efforts reinforced social and economic inequalities, prioritized suburban growth, and supported a plan of industrialization that was largely ruinous for nonwhite communities.

The Failure to Deliver:
Single-Member Districts and Master Planning, 1970–2000

In the 1970s, more and more members of Austin's City Council were being elected with the support of interest groups that were made up of sizeable concentrations of students, Hispanics, African Americans, neighborhood associations, or environmentalists, and these City Council members increasingly opposed many of the progrowth priorities advocated by the business community's principal organizations. While these interest groups were never completely aligned in a long-term ruling coalition, from time to time over the next several decades, they formed a loosely organized majority voting bloc on the City Council. Two of the central priorities advocated by this coalition were changing Austin's system for electing local City Council members to a district system and adjusting the priorities in the city's comprehensive plan to reflect a vision of growth management.

An important development that affected the composition of the City Council at this time was the emergence of students as a significant electoral force. The student population at the University of Texas grew dramatically in the 1960s, as did the percentage of students in Austin's population: in 1964 enrollment was about 26,000 (about 12 percent) of the city's population; by 1973 it had risen to over 40,000 (around 16 percent) (City of Austin 1974, 20). Realizing the potential strength of this demographic reality and the future impact of the 1971 constitutional amendment lowering the national voting age to eighteen, a group of students organized a voter registration effort and a campaign to elect a City

Council member in 1970. Largely as a result of these efforts, in 1971 Jeff Fried-
man, a twenty-six-year-old, won a seat on the City Council; four years later he
became the city's second popularly elected at-large mayor (Orum 1987, 272–273).
However, because Austin has a weak-mayor system, the mayor has no more executive power than other City Council members.

Another factor that significantly influenced local politics was the rising
power of Hispanics as a more or less coherent political group. In 1950, it was
estimated that Hispanics were about 10% of the city's population; by 1970 this
number had increased to about 15% (City of Austin 1974, 14). While Hispanics
had faced discrimination in Texas since the nineteenth century, until the 1950s
their main strategy to gain equal treatment had been to demand their legal
standing as "whites"; under Texas law and court rulings, all persons not "black"
(without "African ancestry") were legally white (Behnken 2011). However, these
efforts proved largely ineffective at lessening the pervasive unequal status of
Hispanics (Macdonald 2012). In fact, after the 1950s, "real white people" used
Hispanics' legal status as "whites" to preserve their own privileges. For instance,
in the 1960s Austin's school authority tried to integrate the city's school by sub-
terfuge, contending that since Hispanics were counted as "whites," school integration could be accomplished without busing because Hispanics could simply
be placed into schools with high African American student populations (Davis 1975, 54–55). While Hispanics remained largely "white" by law, increasingly
some court rulings, beginning in the 1950s, affirmed their legal standing as a
distinct group within Texas that was subjected to illegal forms of discrimination
(Martinez 1993). Furthermore, in the 1960s, the unique position of Hispanics
within Texas's racial hierarchy became the point of mobilization in the growing
Chicano movement, a group that, instead of demanding to be treated as "white,"
claimed recognition as a distinct cultural group (Montejano 2010, 263–270).

A 1967 charter amendment, adopted by referendum, created two additional
at-large places on the City Council and had significant implications for shatter-
ing the color line for local elected officials in Austin. In 1971 African American
Berl Handcox won a seat on the City Council, making him the first African
American City Council member since Reconstruction. Four years later John
Trevino became the first-ever Mexican American City Council member. Some
people have claimed that these successes were the result of a so-called gentleman's agreement struck between the white business community and a group of
conservative nonwhite minorities (Davidson 2012). The agreement, it has been
suggested, created these two places for nonwhite candidates who were accept-
able to this dominant power structure and was designed to stave off future legal
challenges to the racially discriminatory effects of Austin's at-large place system
(Largey 2011). Although only circumstantial evidence supports the allegations
of this conspiracy, it is true that on any City Council since 1975, nonwhites have
held at most two of the seven places.

But for many community members, the election of an African American and a Hispanic to the City Council in the 1970s did not sufficiently address the lack of democratic representation. A group of reformers, led by former City Council member Emma Long, began to demand that Austin change from an at-large system to a single-member district system (Moody 1973). These advocates noted that because of historical prejudice and other market and nonmarket forces, Austin's more modestly resourced nonwhite communities resided in highly concentrated numbers in East Austin. In fact, during the 1950s and 1960s, as Austin's population grew, the segregation of nonwhite minorities became more stark (City of Austin 1979). In particular, the growing population of Hispanics was settling in a large but territorially contiguous area of East Austin (City of Austin 1974, 14). Advocates contended that a district system might allow nonwhite minorities to become more easily elected to the City Council, which in turn would more than likely increase minority representation on the City Council, especially for Hispanics.

Attempts to challenge the at-large system of representation found in many local governments were common in Texas in the 1970s and 1980s, and provisions in the 1964 Federal Voting Rights Act (VRA) meant to protect minority voting supported these efforts. Beginning in the 1970s, a coalition led by Hispanic and African American groups used these VRA clauses to successfully challenge the at-large systems in local governments around the Dallas metropolitan area. Although courts ruled that this system of representation had effectively limited the power of minorities to elect their own candidates, their rulings maintained a very high threshold for showing the discriminatory effects of at-large systems (Davidson and Korbel 1981, 984). In 1975, an amendment to the VRA broadened the class of persons protected from voting discrimination from only African Americans to include, among other groups, Hispanics. The Mexican American Legal Defense and Education Fund (MALDEF) had been a great advocate for this change because of a concern about San Antonio's liberal annexation efforts and their effect on Hispanic representation in the city's at-large districts. Under the new guidelines, all redistricting efforts in Texas, including those of local governments, became subject to Federal Justice Department approval. According to one study, this provision was "instrumental in San Antonio's change from an at-large to a pure single-member district plan in 1977" (Brischetto et al. 1994, 246). However, U.S. Supreme Court rulings continued to affirm a high standard to show how at-large systems discriminated against nonwhites. As a result, in 1982, supporters helped pass an amendment to the VRA that changed the standard to prove discrimination in voting from "discriminatory intent" to "discriminatory effect" (Baiz 1983). Following this new standard, "minority vote dilution"—that is, any practice that reduced the chances that minorities could elect preferred candidates by pooling them with other groups—became a more common legal challenge to at-large systems (Davidson 1992).

Despite the victories reformers had across Texas, in Austin from 1973 to 2011, every attempt to abolish the at-large system of representation and replace it with a district or district/at-large hybrid system failed (see figure 13). In the first unsuccessful attempt at reform in 1973, a referendum garnered about 40 percent support among the electorate. In response, the NAACP and MALDEF filed a suit in federal court in 1976, but in 1977 a judge ruled against these plaintiffs and held that because a "black and Mexican-American" had been elected at-large to the City Council there was "evidence" that their "interests" were being represented (United States Court of Appeals Fifth Circuit 1984, quoted in Volma Decision, 957). The following year, in 1978, another ballot proposal to create a district system did not garner even 30 percent approval from the electorate. Undeterred, supporters filed another lawsuit in federal court in 1984 (in the wake of the 1982 amendments to the VRA) to force the city to adopt a single-member district system; while the City Council agreed to settle and implement a district system, a judge refused to compel the City Council to accept the agreement (Tucci 1984; United States Court of Appeals Fifth Circuit 1984). The next year, an identical reform measure was put before the electorate, and with about the same turnout, support for the measure was just over 40 percent. Although court challenges stopped, reformers continued to press on, but local referendum efforts in 1988, 1994, and 2002 also faltered (Dunbar 2011). In response, reformers turned to the state legislators that represented Austin's suburbs, most notably Republican state senator Jeff Wentworth, who began to introduce bills that would force Austin to use a single-member district system (Coppola 2011).

From 1970 to 1994, supporters and opponents of reforms to the at-large system in Austin tended to form into more or less well-defined camps. On one side, students, nonwhite minority communities, nonwhite minority City Council members, and especially members of Austin's neighborhood associations advocated for reform. They contended that in addition to disenfranchising nonwhites, the at-large system undermined the accountability of City Council members to neighborhoods and favored the election of conservative and wealthier candidates who were friendly to the priorities of the business community's principal organizations. On the other side, opponents, like the progressive reformers from the early twentieth century, argued that the at-large system elected better representatives who were more interested in citywide interests rather than extreme candidates or those interested only in the condition of individual districts. These opponents were mainly drawn from the more conservative wing of the City Council and the business community, particularly the Chamber of Commerce (BeSaw 1973; Tucci 1985; Wright 1994). In the mid-1980s, the fight was particularly intense as the opposition became fiercer, in part because neighborhood associations and nonwhite minorities envisioned redistricting as an essential way to advance their core priorities (Tucci 1984).

For many members of the City Council elected in the 1970s, a central priority was also to use comprehensive planning to manage Austin's largely uncontrolled growth. This goal was supported by a charter amendment passed in a 1973 referendum that removed the requirement that Planning Commission members own real property and also mandated that at least four of its nine members *not* be "directly or indirectly connected to real estate or land development" (City of Austin 1973). Beginning in 1973, under the authority granted to it in 1953, a newly composed Planning Commission, with the help of Austin's Planning Department, began to rewrite the city's master plan. Intentionally involving much more citizen participation than previous comprehensive plans, this planning effort, while largely guided by input from a select group of "neighborhood representatives," allowed hundreds of people to become actively involved in the planning process (City of Austin 1980, 7–8). The result of this effort was the Austin Tomorrow Plan, released in 1977 and adopted in its entirety by the City Council in 1979.

The Austin Tomorrow Plan was unique because in contrast to previous master plans it largely reflected the aspirations of those "Austin citizens [that were] concerned about the destructive effects of continued urbanization on their neighborhoods and natural environment" (City of Austin 1980, 9). So unlike earlier comprehensive plans that envisioned urban planning as a means to promote urban growth, the goal of the 1979 plan was growth management. The plan's recommendations were predominately guided by an overall vision that foresaw the majority of new urban growth being channeled and contained in a narrow but elongated area that followed the path of the Interstate 35 corridor. The primary mechanism proposed to influence the pattern of urbanization was to limit the city's support for utilities and other capital improvements outside this area.

The process of writing and adopting the 1979 Austin Tomorrow master plan ended up being enormously contentious, and the plan was vociferously resisted by the business community's principal associations, so the fact that it was adopted was an impressive achievement for its supporters. In the long run, however, its framework for growth management proved largely ineffective. Butler and Myers argued that at least three factors undermined the plan's efficacy: (1) in 1975 Austin's electorate, divided about the potential effects that its related bonds might have on the implementation of the plan's vision, began voting against the funding mechanisms meant to "provide the necessary funds for capital improvements" and "provide the infrastructure incentives planned for the northern and southern ends of the preferred growth corridor"; (2) in 1981 the independent water authority, the Lower Colorado River Authority (LCRA), agreed to supply water to developments outside the "preferred growth corridor"; and (3) a 1971 change in Texas law permitted the creation of Municipal Utility Districts (MUDs), new legal entities that allowed developers to issue bonds to support capital improvements required for residential, commercial, and industrial developments and then "levy a surcharge on residents" to "recover the costs" (Butler and Myers 1984, 449–451; Shahin 1980). Moreover, because of the ever expanding network of highways in and around Austin (largely financed by the state government), developments could be located just beyond the city's boundaries and thus be excluded from municipal zoning regulations (Butler and Myers 1984, 456–457). The City Council, while often opposing these MUDs, was forced to accept them as a concession to developers, who in turn agreed to some zoning requirements (Butler and Myers 1984, 452). Moreover, the business community's principal organizations, which had been largely sidelined in the planning process, became more aggressive in vying for political control over the City Council, seeking to defeat some members who had supported the comprehensive plan and pressuring later City Council members to not follow its recommendations (Swearingen 2010, 120–122).

 To address the ineffectiveness of the Austin Tomorrow comprehensive plan, beginning in 1986 an attempt was made to update the plan and make it conform to a new requirement, passed by the electorate the year prior, that established "comprehensive planning as a continuous and ongoing governmental function" (City of Austin 1986). At the time, "Local officials believed that the failure of . . . Austin Tomorrow was due in large part to having excluded certain influential political interests [namely the business community]. The Austinplan Steering Committee therefore [intentionally] represented nine groups: business, culture, environmental, ethnic minority, human services, neighborhoods, public sector, real estate, and community at large" (Beatley, Brower, and Lucy 1994, 187).

 In addition to the ninety-four-person steering committee, task forces were formed around fourteen different priority areas such as transportation and land use, and these groups were to make recommendations to the steering committee. But while the steering committee and the task forces mainly developed the plan's overarching vision, its essence lay in the creation of twenty-two sector-planning areas of the city (City of Austin 1988). Large-scale citizen participation was vital to creating maps and recommendations for future land uses, economic activities, transportation improvements, and other functions in each sector.

 In the end, however, the comprehensive planning effort failed to gather sustained public support. It did offer significant amendments to the 1979 master plan, envisioning potential urban growth in a much larger area and planning for much more road construction in both the central city and the surrounding suburbs. Moreover, it proposed the creation of a bike-thoroughfare network throughout Austin and encouraged more development concentrated in central nodes that would be connected by mass transit. Nevertheless, as time wore on, the planning process "became riddled with competing interests[,] and its consensus based approach collapsed as infighting prevented individuals and groups from reaching agreement" (Lopez 2006, 61). At the time, newspaper accounts attributed its failure to attempts to do too much, lack of coherence of dedicated groups, and strong opposition from the business community (Pope 1989; Sargent 1989). Moreover, the master plan's most ambitious aspect, which was that it would become a legally binding document governing the city's growth, was nullified by the state legislature's passage of a bill, led by Terrell Smith, that made all comprehensive plans non–legally enforceable guidelines (Texas State Legislature 1989). Finally, Austin became mired in an economic recession that was caused by three factors: the precipitous fall of the price of oil on the world market, the financial insolvency of Savings and Loan Thrifts, and a substantial bust in the local land-development market (see chapter 1).

In the late 1980s, the combination of the land recession, the collapse of the Austinplan effort, and the failure of the Austin Tomorrow Plan to manage suburban growth patterns helped encourage the abandonment of comprehensive planning for the next two decades. In the 1990s a system of "neighborhood planning" emerged. Largely due to the efforts of council member Jackie Goodman, a twenty-four-member Citizen's Planning Committee (CPC), chaired by architect Ben Heimsath, was created in 1994 to develop a new planning framework for Austin, one largely based on the principles of New Urbanism and designed to bring the business community back into the planning process. Even before the City Council embraced these principles for urban development, this group pioneered collaborations with the business community in the mid-1990s to promote a New Urbanist vision among Austin's governing bodies (Austin American-Statesman 1995a; Phenix 1996). Between 1994 and 1997, several reports released by the CPC, reflecting the influence of New Urbanism, recommended redeveloping the central business district and East Austin; simplifying and reforming the development code to promote more infill and density; and encouraging more transit-oriented mixed-use projects (Citizens' Planning Committee 1995, 1996, 1999). The business community responded positively to these ideas, especially the idea of rewriting the land-use code (Jayson 1994, Lindell 1995).

However, the centerpiece of the CPC's endeavors was their prescription to transfer planning efforts to the neighborhood scale. Breaking down the city's planning program into smaller units would allow for the creation of a more flexible land-use system, more local land-use variations, and, it was hoped, a more collaborative relationship among developers, neighborhood groups, environmentalists, and the City of Austin's Planning Department (Marban 1995). It would also increase citizen involvement in the planning process (Moscoso 1997). To institutionalize this citizen participation, one of the CPC's central recommendations, later adopted by the City Council in 1997, was to make a registry of existing neighborhood associations, reduce their vast numbers and sometimes conflicting geographical boundaries, and incorporate them as formal stakeholders in the development process (Heimsath 1995). By 1997, after an extensive review undertaken by the city's government, the number of neighborhood associations had been cut in half. Later that year the city began to impose a formal registry system for neighborhood associations, which developers relied on to alert surrounding communities about their future projects (Austin American-Statesman 1997). Advocates of this formal system argued that it was a necessary step in bringing developers back into the planning process because neighborhood protests, which increased a developers' cost by causing delays, were often instigated by a lack of information or misinformation about new developments (Dworin 1995).

Following several pilot projects started in 1996, neighborhood planning quickly spread throughout the city, and by 2012 most development in the city's urban core was guided by a neighborhood plan. Each of these neighborhood plans contained a mix of priorities and preferences of that area's multiple stakeholders, such as developers, residents, and neighborhood associations, which were codified on a future land-use map (FLUM). Like other comprehensive maps, FLUMs are not legally binding as land-use regulations, but once adopted by the City Council, they become amended to the 1979 comprehensive plan and, perforce, do serve "as a guide for future rezoning proposals" (Lopez 2006, 65).

While it was never explicitly stated, at the turn of the twenty-first century it became clear that "the Neighborhood Planning process [was being used] . . . to create a comprehensive plan through the back door"—a use that had several problems (Keeling 2000, 32). First, the charter reform adopted in 1985 mandated that Austin create a workable master plan. Second, not all of the planning areas had adopted neighborhood plans, so there was an uneven patchwork throughout the city's central areas. Third, the areas that had created neighborhood plans were located mainly in the existing urbanized areas. While the majority of people lived in these areas, it was not where the majority of development, which concerned environmentalists, was happening. In fact, urbanization in the outer areas of Austin remained largely unguided even by the 1979 master plan. Finally, the neighborhood planning process had been fraught for both neighborhood associations and developers. On the one hand, while the neighborhood plans engaged neighborhood associations as formal stakeholders in the planning process, the planning areas did not rely on the existing geographical boundaries but rather created new planning districts that sometimes contained several neighborhood associations. As a result, conflict sometimes emerged among associations, particularly over the FLUM's prospective land-use recommendations (McCann 2003; Wilson 2006). On the other hand, members of the business community noted that a "drawback . . . of these neighborhood planning districts is that, very often, existing residents of a community do not want an intensity of use that they feel would compromise neighborhood quality" (Greater Austin Chamber of Commerce 2003, 115).

Furthermore, the New Urbanist planning principles embodied in the neighborhood planning process had been an important factor in the gentrification of East Austin and were opposed by groups that represented nonwhite minorities. Certainly, the CPC's recommendations had encouraged gentrification when it pushed for the redevelopment of East Austin. Perhaps as important, however, is the lingering impact of past public and private zoning in East Austin. One reason it was much easier to build new infill developments in these areas, even if they diverged from the recommendations of a FLUM, is that they lacked many of the private controls over land use that had been adopted in other areas of the city more than half a century earlier.

From the 1970s to the early part of the twenty-first century, the coalition that had managed several times to control the formal levers of power in the municipal government nevertheless failed to reform Austin's system of political representation and develop or implement a workable comprehensive plan. Had reformers of the city's local election system been able to achieve their goal, Austin's nonwhite communities might have been greatly empowered, which in turn might have strengthened the coalition's chances of solidifying its governing position and strengthened City Council's ability to pursue a comprehensive planning vision that reflected the coalition's ideals about how the city ought to grow. Instead, in the late 1990s groups in Austin's fast-developing suburbs, often hostile to mainstream environmentalism, entered into a coalition with the city's nonwhite communities to abolish Austin's at-large system of representation. (This coalition was later strongly supported by the business community.) In comparison, when comprehensive planning was officially abandoned in the 1990s, it was replaced with a system of neighborhood planning that was guided by the principles of New Urbanism. By the late 1990s, this planning paradigm had become the basis for a new governing consensus among the city's mainstream environmental and business communities about the future direction of urban growth: priority should be given to new developments that reorganized development patterns in neighborhoods in the existing urbanized areas. Coincidentally, the communities that were most affected by this new planning agenda, more modestly resourced communities living in the existing urbanized area, were not its leading supporters.

Imagining a Better Austin Tomorrow, Today

The priorities of the governing coalition that was born in the 1990s are reflected in the Imagine Austin master plan, which, like nearly all of the city's planning efforts over the last two decades, envisions a city that has a denser, more connected urbanized area. The expectation among its supporters is that this will create the conditions for a more livable city with vibrant communities and a healthy natural environment, although there is evidence that the benefits associated with this new form of urbanization, as chapters 4 and 5 showed, may not be available to many of Austin's more modestly resourced communities (Mueller and Dooling 2011). Moreover, nonwhite minorities were largely excluded from the comprehensive planning process, not by intention but by omission. A recent study on public input for the Imagine Austin plan shows that it mainly came from well-educated white people living in the central city, findings that are very similar to a 1987 study of public participation during the debate over the Austinplan (Goodspeed 2010). Therefore, the input gathered from the "public" that was used to justify the priorities of the Imagine Austin plan may merely reflect the leading priorities of the members of the governing coalition—that is, key representatives from the business and environmental communities.

But is there a connection between the support of major business community association for single-member districts and the comprehensive planning effort like there was in the past? While few public documents account for the business community's motivation for supporting the political reform, a 2003 audit by an independent consulting firm commissioned by the Chamber of Commerce on Austin's "business climate" does suggest a connection to urban development. The report notes in the executive summary section on "business costs" that "the regulatory and permitting processes in the city of Austin are time consuming, arbitrary and cumbersome. An at-large City Council with no district representation, numerous, powerful neighborhood groups and an active environmental lobby in Austin make the process of receiving permit approval often difficult" (Greater Austin Chamber of Commerce 2003, 7).

The aforementioned statement is especially interesting because it was based on "numerous individual interviews and focus groups" with leading members of Austin's business community as well as "an extensive online survey" of 174 respondents (Greater Austin Chamber of Commerce 2003, 1).Thus, it provides an excellent window into the attitude of leading members of the business community toward the electoral reform. Later in the same report, the connection is stated more baldly: "Because Austin's councilpersons are elected at-large, they have no direct stake in the particular neighborhoods in question, and therefore may impose citywide views on neighborhood-area issues" (Greater Austin Chamber of Commerce 2003, 114). This section of the report is referring to the neighborhood planning process and its problems, particularly the ability of an individual property owner to delay the formation of zoning districts. Nevertheless, in context, the statement suggests that despite the tremendous amount of development in Austin over the previous thirty years, the business community's representatives still believed that the at-large system favored the creation of a "Byzantine" land-use code and complex FLUMs that empowered environmentalists and neighborhood associations and produced unnecessary delays in permitting and zoning.

What about the Imagine Austin Plan? Will it also help relieve the problems that were identified as the main priorities of the leading business groups? Certainly the passage of the comprehensive plan has initiated a complete rewriting of the land-use code, something that had been opposed by both environmentalists and neighborhood association activists for decades, and something they opposed in their public comments on the Imagine Austin Plan (Stone 2012). Moreover, there will be discrepancies between the master and neighborhood plans, especially if the density priorities proposed in the comprehensive plan do not align with the development goals set forth in individual neighborhood plans. At this point it is unclear which of the two plans will play a greater role in guiding future decisions about variances in land use and zoning, but the master plan does provide a justification for modifications, particularly for greater density in the existing urbanized areas of the city. However, the key to the support of the business community's principal organizations for the comprehensive plan, as it was with past plans, may lie not in how the plan affects day-to-day planning decisions but rather in how its vision guides priorities for future bonding. A defining feature of the Imagine Austin plan is its emphasis on rescaling the city's planning efforts to the region, a priority that is overwhelmingly supported by the principal business associations. Echoing the priorities identified in strategic documents of the business community's principal representatives, the plan stresses the need for a more regionally integrated transportation planning and governance. The rescaling of Austin's planning efforts may in fact be the most significant aspect of the plan because this geographical reimagining will guide priorities for future capital improvements, particularly the spending of millions of dollars to expand the city's transportation network to create a more effectively interconnected region.

Outside the Shadows

If democracy is to mean anything, it is the ability to all agree to arrange things in a different way.

David Graeber, *Debt: The First Five Thousand Years*

Repetition

In 2013, popular *Texas Monthly* magazine published a collection of commentaries on Texas's major cities, and two of them were about Austin. One, written by John Spong (a *Texas Monthly* staff reporter), titled "All Grown Up," suggested that although Austin's physical and social environment had changed substantially over the previous twenty years, the city still retained an abiding culture of tolerance to difference, a culture that not only provided the city with its great quality of life but was the main factor in propelling its phenomenal growth. While reserving animus for those who complained that the city's growth had changed it for the worse, Spong amicably parroted the tall tale (that has been repeated ad nauseam) about the large role the city's culture played in driving its rapid urban development. Notably, his article attracted very little commentary, critical or otherwise.

The other commentary, penned by Cecilia Balli (an associate professor of journalism at UT Austin), was titled "What Nobody Says about Austin." In it she declared that Austin was the most segregated city in Texas and that its citizens refused to confront the lingering impact of white supremacy on the city's governance. But perhaps most importantly (at least in terms of the public response), she insinuated that while Austinites were supposedly renowned for their tolerance to difference, their open-mindedness was, by and large, extended only to the quirky behavior associated primarily with "white" people. Balli's article prompted a vigorous discussion in Austin's blogosphere, and while there was some support for her editorial, it was overshadowed by a torrent of outrage.

Although Balli can be faulted for some factual errors, ahistorical assumptions, and hyperbolic claims, the vast majority of the negative responses to her article did not point to any of these issues. Instead, they relied on racist, misogynist, and other ad hominem attacks that, more than anything, showed how intolerant Austinites can be, especially to those who cast even a small amount of shade over the city's excessively sunny image.

The articles by Spong and Balli poignantly represent two of the most common conversations taking place in and about Austin today. On the one hand, there is an interminable debate about whether the changes in the city's physical infrastructure are good, principally as they pertain to Austin's cultural life. On the other hand, there is an ongoing discussion about Austin's stark pattern of racial segregation and how what is considered "Austin's culture" is pregnant with notions of whiteness. Both of the authors, despite speaking to only one of these discussions, are optimistic that the problem they identify can be solved. One suggests that there is no cause for alarm because of the durability of a local sensibility, and the other, while concerned, believes that a new multicultural sensitivity to difference will emerge. However, both authors overlook the more fundamental question driving both of their discussions: who has relatively more control over who (or what) bears the costs associated with how the city has been developing?

In this book, I have demonstrated how the past and present patterns of urban development in Austin have been strongly influenced by a historically varying, but still relatively stable, system of asymmetrical power relations that has engendered both uneven development among neighborhoods and inequalities among peoples. From urban renewal programs to Smart Growth planning efforts, those people who were in a relatively more vulnerable position than others (for a variety of reasons) have shouldered more of the costs associated with changes to the city's urban environment. Moreover, a series of growth coalitions with different, albeit similar, political configurations have had determinately more influence over local governance and the policies and programs that were (or could be) enacted to facilitate specific changes in Austin's urban form. If there is any hope of resolving the issues brought up in these popular discussions, the discussion has to recognize the significant role of social, geographical, and institutional unevenness, both historically and contemporarily, in shaping the varying fortunes and fates of different people and places.

Summary

In the first three chapters I pointed out how the dynamic industrial and urban development in Austin is significantly connected to the ability of the city's growth coalition to take advantage of two interrelated factors: (1) the restructuring of the national economy, where the leading industrial sectors shifted from activities involving labor-intensive manufacturing toward knowledge-intensive services; and (2) the changing geographical patterns of industrial investment, which supported the urbanization of new regions such as the Sunbelt. Moreover, I noted how Austin's recent growth dynamics might have benefited from a number of preexisting local factors such as the presence of a large university, a relatively educated population, a strong institutional relationship among members of the growth coalition, and a racially stratified workforce dominated by services.

However, I departed significantly from previous studies on Austin by arguing that the city's urban geography, both its physical morphology and its political structures, still had to be considerably modified in order for Austin's growth coalition to take advantage of the changing value of knowledge production and dissemination as well as the shifting national and international flows of investment (in terms of both sector and location). To be precise, in chapters 2 and 3, I showed how UT Austin was an essential actor in facilitating a significant spatial, and consequentially a social and economic, transformation of the city. Certainly, the university's role as a knowledge factory, first in producing a highly technically skilled workforce and later as the holder of intellectual patents, helped produce local businesses and change the region's geography. But, as I stressed, the university also had a potent and unique role as a major land developer, and in this capacity it substantially influenced how Austin's urban geography was made or remade to support the growth coalition's industrial development strategy and advance the city's firm orientation toward what is now called the knowledge economy. Moreover, and in part as a consequence of the university's land development efforts, the configuration of the local growth coalition was significantly altered. In particular, the university's administration, and by extension the state government, became much more entrenched as a leading actor within the local growth coalition.

Furthermore, I contended (greatly diverging from the existing literature on Austin in the process) that the terminology developed by David Harvey in his model of the expanding reproduction of capital could account for the urban growth dynamics revealed in this case study. First, I argued that how capital flows into and out of three possible circuits is a significant factor in shaping patterns of urbanization because capital switching among circuits affects not only sectors but also, as a result, places. Second, I argued that capital flows into secondary and tertiary circuits and the associated appropriation of revenue from technological and knowledge rents has a substantial role in driving contemporary patterns of urbanization, especially in highly industrialized economies. However, analyzing Austin's urban development through this approach also revealed how the city may be an exceptional case, a place where, as a result of the lingering influence of past development strategies and the city's less elevated position within the global urban hierarchy (a reflection of its intense specialization in certain producer services), the revenues from knowledge and technological rents have come to play an unbalanced role as drivers of the city's growth.

The overarching theme that runs through the book's last three chapters is connecting two issues: (1) the effects on urban life in Austin from implementing various planning regimes; and (2) the durable significance of the business community in shaping the city's urban planning policies and priorities. The main point evinced by these chapters is that today, as in various periods in the twentieth century, the business community's principal organizations have a disproportionate influence on how urban planning policies and priorities take shape. In particular I focused on how the widespread embracing of several new urban design principles, principles that continue to play a very strong role in informing present planning practice, resulted in part from the business community's willingness to reach a political compromise with a leading political faction of the environmental community. As in previous times, this compromise has had significant consequences for communities not included in this governing coalition. While there may not be anything particularly special about the business community's unequal influence in Austin's local governance, especially in Texas, the effects on the city's development are unique. In particular, Austin's planning efforts have had distinctive local effects that are reflected in the city's urban form.

Synthesis

There seems to be an association between cities that, like Austin, are both leading centers in the knowledge economy and leaders in meeting sustainable development goals, especially as they pertain to the environment. But is there a substantial connection between the contemporary economic processes (characterized by distinctive forms of rent seeking) that innervate Austin's industrial and urban development and the particular planning policies and priorities (broadly covered under the rubric of urban sustainability) that have emerged to reorganize the city's urban form?

Richard Florida's famed argument about the role of the creative class in shaping urban development patterns would suggest a connection. In a nutshell, his argument is that competitive cities remain competitive because of their ability to capture and retain certain kinds of highly technically skilled (talented) workers and other so-called creatives. And because one strong preference of such workers is to live in places with lots of ecological amenities, urban growth coalitions support policies attuned to sustainability, particularly in terms of the natural environment. Here is how Florida describes the connection in Austin in his hit book *Cities and the Creative Class*:

Austin, Texas, has developed a two-pronged strategy for its economic future: high technology and smart growth along with lifestyle amenities. In 1998, the Greater Austin Chamber of Commerce undertook a new regional strategy outlined in a report entitled Next Century Economy. . . . The report identified three strategies for the city to follow in developing sustainable advantage. The first strategy involved bolstering the region's already thriving high-technology economy by (1) improving communication between the region's economic clusters and their suppliers, (2) improving the economic foundations of existing businesses, (3) leveraging clusterbased R&D at the University of Texas, and (4) attracting firms to complement existing industries by filling supply gaps in existing clusters. The second strategy called for the region to "ensure environmental quality and social opportunity by explicitly linking social and environmental goals to economic development goals." . . . The third strategy involved developing a regional collaborative mechanism for major problem-solving, in particular for linking high-technology development to smart growth and amenities. . . . The Austin case illustrates how far-sighted regions are recognizing that continued success in the high-technology economy will turn on the ability to deliver environmental quality, natural amenities, and the lifestyle desired by knowledge workers. (Florida 2005, 66–67)

So, according to Florida, the competitive race against other cities for more members of the so-called creative class is the reason why Austin's growth coalition shifted in the late 1990s to encouraging the City Council to make strategic investments that would increase the city's stock of social and ecological capital, especially by using its powers over urban planning.

Contra Florida, a body of scholarship has emerged that argues, in varying ways, that urban sustainability and urban competitiveness have not converged. Instead, the priorities associated with improving environmental quality or social opportunity are only selectively incorporated into an urban growth agenda. In particular, these principles are used in the innovative redesigning and upgrading of areas of a city that have been relatively underdeveloped, but the goal is to make them more attractive to middle-class professionals. Florida is often the target of criticism because he has supported a broad range of redevelopment programs, such as gentrification, that have been shown to exacerbate regional inequalities. Yet nearly all of his detractors concede that the primary factor driving cities to transform their urban forms and planning regimes is the competitive struggle to capture the small pool of knowledge economy elites—that is, young, highly educated, mobile, and (primarily) white people. Here is how Eugene McCann, in an article about Austin, put it:

Contemporary urban policy-making aimed at nurturing, attracting, and retaining the group of workers and capitalists Richard Florida dubs the Creative Class has partly entailed the deployment of a particular geographical imagination. This imagination is underpinned by an ideal of vibrant, creative urban neighborhoods and by a mental map of cities to be learned from in terms of good urban policy and to be competed against for creative talent and high-tech investment. This geographical imagination and its intrinsic spatial frames are important, I suggest, because they are causal stories that encourage and legitimate specific policy interventions in the built environment and in the economic base of cities. (McCann 2008, 9)

Like Florida, then, McCann argues that the principal reason Austin's growth coalition supported the development of a new urban planning framework for the city was the framework's ability to attract and retain members of the so-called creative class.

Although I am sympathetic to McCann's interpretation (and make a similar argument elsewhere in this book), I wager that a different account can be offered for a probable connection between Austin's ability to continue to garner private investment, related to the knowledge economy, and the business community's willingness to link social and environmental goals with economic development. Essentially, I propose that the business community supported planning programs and goals designed to help leading knowledge-intensive firms in Austin, including the university, gain a competitive advantage from the temporal benefits that will flow from the massive investments to refashion the city's urban form. Any environmental or social benefits of this planning regime may be serendipitous. What this means is that the business community's leading organizations did not have to consider social or ecological benefits when supporting the city's new dominant urban planning framework and its attempt to engineer a new spatial and temporal configuration.

Today, the pressures of time and speed, as they relate to the ease of coordination, are essential for the competitive advantage of highly knowledge-intensive firms (Sassen 2011). As a consequence, spatial considerations are vital because despite the ability of communication technologies to more speedily link places together, some temporal benefits can be realized only when there is considerable spatial centralization—for instance, in a specific office building, a central business district, or a specialized research park. However, as in the past, firms face a significant spatiotemporal dilemma because access to the places that may give a firm the greatest temporal advantages comes at a very high financial cost, as landowners, understanding these benefits, demand a high rent for use of this prime real estate. Hence, some firms choose to locate in a less-than-prime place, often in an agglomeration of firms concentrated in a marginal area, because the monetary savings in rent may be greater than the competitive advantages that could be gained from being in a better location. However, the relative differences among locations can be diminished by urban planning efforts that enhance the connectivity of a polynucleated region and bring its central nodes into better communication with one another. On the one hand, this would help diminish the temporal disadvantages that some firms suffer as a result of their location. On the other hand, it would assist in locking in future high real estate values in existing centers of development because of the massive physical infrastructure investments that are made to link these areas more easily to one another. Moreover, promoting temporal efficiency and spatial cohesion would produce significant ecological and social benefits, particularly if infrastructural improvements were made in the form of better mass transit, more connected and denser housing, and an enhanced information and telecommunications network.

Additionally, the more connected and densely organized a city's urban form becomes, the more its university (or group of universities) becomes embedded in the local network of knowledge-intensive firms. In turn, the competitiveness of the regional innovation system will be improved because ties between the university and local firms will accelerate the transfer of knowledge from the university to the firms and, therefore, facilitate the creation of new knowledge-based commodities (Cooke 2005). Moreover, saving time pays huge dividends in the knowledge economy for at least two additional reasons. First, substantial returns from technological or knowledge rents require scientific discoveries that are then transformed into tradable goods and disseminated as commercially viable products. However, science advances only as a collective social enterprise, and while new research breakthroughs can be made at tremendous private or individual effort and expense, similar discoveries are often made simultaneously in many places. Thus, being the first person (or group) to make the claim on a new discovery is paramount. Second, scientific discoveries, and in fact most information that is pressed into knowledge-based products, are by their very nature nonrivalrous (their use by one person does not preclude their use by another, nor is their quality denuded if others use them). While the generalized dissemination of knowledge can be restricted by enforcing legally granted monopolies or by keeping the new findings a secret, once knowledge is shared (by theft, accident, or some other means), it can be easily copied and employed by anybody, threatening its potential commercial value. Thus, there is an urgency in patenting new knowledge as quickly as possible, bringing original discoveries to market very rapidly, and pushing firms to adopt them immediately. Therefore, in order to save time, it is crucial for the university to be in close proximity, or at least connected as seamlessly as possible, to a concentration of other knowledge-intensive firms and other specialized producer services that can turn scientific discoveries into profitable streams of revenue.

While the account I have given above is rather general, let us now turn to the specific vision of growth and Austin's future urban form found in the transit plan developed in the city over about the last decade, in order to get a sense of the relationship between the regional Smart Growth (now urban sustainability) planning goals and the knowledge economy and to understand why the local growth coalition has so enthusiastically endorsed this plan. First, let us look at Austin's first commuter train, which opened in 2010 and is an essential part of this plan. Although the train's route is strange because it was forced to share a preexisting train track, it connects the entertainment district (located in the central business district) and the rapidly gentrifying neighborhoods in East Austin to some of the fastest-growing nodes of development for high-technology firms in the northwestern suburbs. Perhaps more telling is the vision found in the more comprehensive program for mass transit, which was embodied in Project Connect, the failed transportation bond initiative of 2014. Certainly the business community had a lot to like in that specific bond package because it contained two implicit development priorities: (1) it linked the financing for only a small section of the proposed urban mass transit system to a group of contentious new road projects in Austin's suburbs; (2) it prioritized future redevelopment goals, particularly the prospective increases in population density and land values that would occur along the proposed rail line, over the needs of existing communities that were the most likely to be its users. Moreover, one of the explicit priorities of the proposed mass transit system was to provide a new link to the main campus of the University of Texas. However, the proposed rail line would have done little to help the thousands of undergraduates and employees commuting to and from the university because its projected route was along the remote eastern edge of the main campus. Instead, the main beneficiaries of the new mass-transit connection would be the proposed biotechnology research hub as well as the new medical center and the innovation district of knowledge-based firms that are being built just east and south of the university. Furthermore, it would have connected these new areas of growth to the burgeoning residential redevelopments found in the East Riverside Corridor and the Mueller Community; the regional airport; and the advanced producer services, such as commercial banks, financial firms, and specialized law offices, located in the central business district.

The interpretation I have provided shows that it is possible to account for the agreement between contemporary economic processes and efforts to rework Austin's urban form without offering that the city's business community has embraced (or not embraced) a stronger commitment to the goals associated with urban sustainability. To this extent, the literature's focus on individual preferences appears overstated: modest improvements in regional transportation efficiency and greater housing density, regardless of intention, will have some selective ecological benefits. Should future urban research move its focus back to the behavior of firms? No! Instead, the question is whether urban planning should always be subordinated to the enhancement of a city's or region's competitive position. I now turn toward this question.

Future

I ended the first chapter of this book by suggesting that many attempts to create a different kind of Austin have been foreclosed by a mixture of local and global circumstances, but I also suggested that understanding how these forces have shaped development in Austin could open up new opportunities for re-imagining the city's future. I even went so far as to suggest that revealing these processes might allow us to advocate for different, possibly better, forms of regional development that would invest in the quality of development rather than its quantity.

The results of these approximately fifteen years of sustainable urbanism in Austin have been mixed. Despite some laudable efforts, between 1990 and 2010, Austin's urbanized land area grew by 150 percent, and the population of its "never-urbanized" area increased by about 160 percent; as a result, the density of urbanized areas in the city decreased by approximately 15 percent (Lewis, Knaap, and Schindewolf 2012). Moreover, over the same period, there were some interesting trends in the Austin MSA's Gini Coefficient (a statistical measurement used to capture how much the distribution of income within an economy at some fixed spatial scale deviates from a perfectly equal distribution). Although the coefficient for the entire MSA decreased by 0.061 (to 0.780), when only the urbanized area is measured, the coefficient rose by 0.010 (to 0.360). A coefficient closer to 1 represents greater inequality, so these changes suggest that while income was deconcentrated across the region, probably because of rapid population growth in the suburbs, a few people within the existing urbanized areas controlled a greater share of the area's income. A more telling figure, however, might be that from 1999 to 2010 poverty rates in the Austin MSA grew much faster than in other Texas cities, from approximately 14 percent of the population to 21 percent. These figures rose much more dramatically, from circa 21 percent to 30 percent and from 19 percent to 31 percent, for Hispanics and African Americans, respectively (Murdock 2013). Finally, it is worth noting that Austin has the distinction of being the only U.S. city with a population over five hundred thousand that has grown by more than 10 percent over the last decade while losing a portion of its African American population at the same time (Tang and Ren 2014). Many variables have influenced this trend, and it has been taking place for a long time, but it is interesting that the neighborhoods of the city that had the highest concentration of African American residents have been prime targets for recent planning initiatives designed to promote greater urban sustainability.

I have been critical of Smart Growth (or what is now called sustainable urbanism) in this book, but I do think this development paradigm can offer something to our efforts to reimagine and possibly create a better collective urban future. Smart Growth is not a set of values that is reducible to the beliefs and practices of a dominant group of planners or a single interest group. Instead, it is a pliable set of ideals, perhaps too protean for some, that is composed of a vast assortment of competing, often contradictory, value systems (Dierwechter 2008). In Austin, the most dominant variant of Smart Growth is technocratic, has a narrowly focused conceptualization of what counts as an environmental problem, and is strongly aligned with the growth coalition's agenda. For these reasons, there seems to be a persistent problem in meeting sustainability targets associated with the goals of social equity, although meeting environmental goals is also a problem. So there is an urgent, and realizable, task at hand: challenging assumptions about what constitutes sustainable urbanism, recognizing its limitations, and demanding that its goals can be achieved only if democratic control over local and regional planning is expanded. In other words, there may be real opportunities for change in the current state of development, and our task is not just to imagine a brighter urban future but also to bask in the possibilities that confront us in the present.

BIBLIOGRAPHY

Abrams, Charles. 1955. *Forbidden Neighbors: A Study of Prejudice in Housing*. New York: Harper.

Adams, James D. 2002. "Comparative Localization of Academic and Industrial Spillovers." *Journal of Economic Geography* 2(3):253–278.

Agyeman, Julian. 2005. *Sustainable Communities and the Challenge of Environmental Justice*. New York: New York University Press.

Alcalde. 1968. "Focus on the Forty Acres." *Alcalde* 57(1):20–21.

Almanza, Susana. 2010. Personal Interview with former executive director of PODER, February 11.

Almanza, Susana, Sylvia Herrera, and Librado Almanza. 2003. "SMART Growth, Historic Zoning, and Gentrification of East Austin: Continued Relocation of Native People from Their Homeland." Austin: PODER.

Ankrum, Nora. 2010a. "Austin's Political Ecosystem: Scott Swearingen's 'Environmental City' Tells the Inner History of Austin Politics." *Austin Chronicle*, April 23.

———. 2010b. "In-Comprehensible? ANC Balks at Comp Plan." *Austin Chronicle*, December 3.

Arbingast, Stanley, and Robert Ryan. 1982. "Austin in the Eighties." Austin: Chamber of Commerce.

Arnold, Mary. 2008. Personal Interview with community activist, September 18.

Astebro, Tom, and Navid Bazzazian. 2011. "Universities, Entrepreneurship and Local Economic Development." In *Handbook of Research on Entrepreneurship and Regional Development*, 252–333. Cheltenham, Gloucestershire, U.K.: Edward Elgar.

Atkinson, Richard C., and William A. Blanpied. 2008. "Research Universities: Core of the U.S. Science and Technology System." *Technology in Society* 30(1):30–48.

Austin American-Statesman. 1956. "Growth of University Physical Plant Parellels Student, Faculty Increase." *Austin American-Statesman*, August 12.

———. 1995a. "Citizens' Committee Is a Vehicle for City to Coordinate Planning." *Austin American-Statesman*, June 9.

———. 1995b. "Zero Tolerance Falls Short." *Austin American-Statesman*, December 15.

———. 1997. "Neighborhood Plan Flies." *Austin American-Statesman*, May 27.

Austin Area Economic Development Foundation. 1948. "Trade and Civic Characteristics of Austin, Texas." Austin History Center: Austin Economic Development Foundation.

Austin Board of Trade. 1894. *The Industrial Advantages of Austin, Texas, or Austin Up to Date.* http://texashistory.unt.edu/ark:/67531/metapth38097/, accessed December 9, 2013, via University of North Texas Libraries, The Portal to Texas History, http://texashistory.unt.edu. Austin History Center.

Austin Business Journal. 2010. "Real Estate Council of Austin Opposes City Transportation Bond Package." *Austin Business Journal*, October 4.

Austin Chamber of Commerce. 1916. *Progressive Austin.* Austin: Chamber of Commerce.

———. 1961. "Great Academic Stature at UTBlueprinted for 1970." *Austin in Action*, September. Austin: Sharp and Co.

———. 1985. "Creating an Opportunity Economy: Enhancing Quality of Life in a Changing Community: Final Report, SRI Project 7799." SRI International.

Austin Community for Change. 2012. "Specific-Purpose Committee Finance Reports." Austin: Texas Ethics Commission.

Austin Daily Statesman. 1908. "Chartered Discussed." *Austin Daily Statesman.*

Austin Human Relations Commission. 1979. "Housing Patterns Study: Segregation and Discrimination in Austin, Texas." City of Austin. Austin History Center.

Austin Urban Renewal Agency. 1967. Brackenridge Urban Renewal Project Project Number Tex R-94. Austin History Center.

———. 1972. Progress Report '72: A History of Urban Renewal in Austin. Austin History Center.

Austinites for Geographic Representation. 2013. "Specific-Purpose Committee Finance Report." Austin: Texas Ethics Commission.

Austinites for Geographical Representation. 2012a. "Endorsements and Supporters of Citizens Districting 10–1."

———. 2012b. "Specific-Purpose Committee Finance Reports." Austin: Texas Ethics Commission.

Baiz, Dan. 1983. "Hispanics Use New Voting Rights Act to Reshape Texas Politics." *Washington Post*, April 25.

Baldwin, James. 1963. "A Conversation with James Baldwin." In *The Negro and the American Promise*, edited by Kenneth Clark. Boston: WGBH.

Ball, Andrea. 2003. "A New Effort to Help Homeless." *Austin American-Statesman*, October 6.

Banta, Bob. 1994. "Rangers to Boost Downtown Defense against Criminals." *Austin American-Statesman*, March 3.

Barna, Joel Warren. 1992. *The See-Through Years: Creation and Destruction in Texas Architecture and Real Estate, 1981–1991.* Houston: Rice University Press.

———. 2002. "The Rise and Fall of Smart Growth in Austin." *Cite* no. 53 (Spring):22–25.

Barnes, Sarah, and Ted Warren. 1995. "Sidewalks in Need of $51,000 Cleaning." *Austin American-Statesman.*

Barta, Carolyn. 1996. *Bill Clements: Texian to His Toenails.* Austin: Eakin Press.

Beal, Chandra Moira. 2001. "Austin's Downtown Rangers." *Law and Order*, 41–43.

Beatley, Timothy, David J. Brower, and William H. Lucy. 1994. "Representation in Comprehensive Planning: An Analysis of the Austinplan Process." *Journal of the American Planning Association* 60(2):185–796.

Beauregard, Robert A. 1991. "Capital Restructuring and the New Built Environment of

Global Cities: New York and Los Angeles." *International Journal of Urban and Regional Research* 15(1):90–105.

Behnken, Brian. 2011. *Fighting Their Own Battles: Mexican Americans, African Americans, and the Struggle for Civil Rights in Texas.* Chapel Hill: University of North Carolina Press.

Belina, Bernd, and Gesa Helms. 2003. "Zero Tolerance for the Industrial Past and Other Threats." *Urban Studies* 40(9):1845–1867.

Benningfield, Damond. 1988. "Sharpening Austin's Competitive Edge." *Alcalde*, 19.

———. 1989. "New Chips on the Block." *Alcalde* 77(3):26–28.

Berman, Elizabeth Popp. 2011. *Creating the Market University: How Academic Science Became an Economic Engine.* Princeton, N.J.: Princeton University Press.

Berman, Evan M. 1990. "The Economic Impact of Industry-Funded University R&D." *Research Policy* 19(4):349–355.

BeSaw, Larry. 1973. "Charter Proposal Opponents Accused of 'Misstating Facts.'" *American Statesman*, April 3.

Bohmfalk, James Winton. 1968. "The Austin Chamber of Commerce: A History of the Organization and Its Uses of Propaganda." MA thesis, University of Texas at Austin.

Bowie, Norman E. 1994. *University-Business Partnerships: An Assessment.* Issues in Academic Ethics. Lantham: Rowan & Littlefield.

Braudel, Fernand. 1982. *Civilization and Capitalism, 15th–18th Century: The Perspective of the World.* Berkeley: University of California Press.

Brenner, Robert. 2003. *The Boom and the Bubble: The U.S. in the World Economy.* London: Verso.

Brewer, Anita. 1965. "Where Are We Going Next? The Expanding Boundaries of the Forty Acres Move North, South, West, and Mostly East." *Alcalde* 54(2):16–19.

Breyer, R. Michelle. 1997. "Downtown Group Set '98 Agenda." *Austin American-Statesman*, October 10.

Bridges, Amy. 1997. *Morning Glories: Municipal Reform in the Southwest.* Princeton, N.J.: Princeton University Press.

Brint, Steven. 2005. "Creating the Future: 'New Directions' in American Research Universities." *Minerva* 43(1):23–50.

Brischetto, Robert, David P. Richards, Chandler Davidson, and Bernard Grofman. 1994. "Texas." In *Quiet Revolution in the South: The Impact of the Voting Rights Act, 1965–1990*, edited by Chandler Davidson and Bernard Grofman. Princeton, N.J.: Princeton University Press.

Brooke, Ann. 1980. "Plugging into Austin's Booming Electronics Industry." *Austin: An Official Publication of the Austin Chamber of Commerce*, October, 39–47.

Brown, Norman D. 1984. *Hood, Bonnet, and Little Brown Jug: Texas Politics, 1921–1928.* College Station: Texas A&M University Press.

Brownell, Blaine A. 1975. *The Urban Ethos in the South, 1920–1930.* Baton Rouge: Louisiana State University Press.

Browning, Larry, and Judy Shetler. 2000. *Sematech: Saving the U.S. Semiconductor Industry.* College Station: Texas A&M University Press.

Brubacher, John Seiler, and Willis Rudy. 1997. *Higher Education in Transition: A History of American Colleges and Universities.* New Brunswick, N.J.: Transaction.

Bullard, Robert D. 2007. "Introduction." In *Growing Smarter: Achieving Livable Commu-*

nities, Environmental Justice, and Regional Equity, edited by Robert D. Bullard, 1–19. Cambridge, Mass.: MIT Press.

Bullard, Robert D., and Glenn S. Johnson. 2000. "Environmental Justice." *Journal of Social Issues* 56(3):555–578.

Bunch, William. 2008. Personal Interview with executive director of Save Our Springs, September 4.

Burger, R. M. 1979. *An Analysis of the National Science Foundation's Innovation Centers Experiment*. Washington, D.C.: United States Government Printing Office.

Busch, Andrew. 2011. "Entrepreneurial City: Race, the Environment, and Growth in Austin, Texas, 1945–2011." PhD diss., University of Texas at Austin.

———. 2013. "Building 'a City of Upper-Middle Class Citizens': Labor Markets, Segregation, and Growth in Austin, Texas, 1950–1973." *Journal of Urban History* 39(5): 975–996

Butler, John Sibley. 2004. "The Science and Practice of New Business Ventures: Wealth Creation and Prosperity through Entrepreneurship Growth and Renewal." In *Entrepreneurship: The Way Ahead*, edited by Harold P. Welsch, 43–54. London: Routledge.

———. 2010. "The University of Texas at Austin." In *The Development of University-Based Entrepreneurship Ecosystems: Global Practices*, edited by Michael Fetters, Patricia G. Greene and Mark P. Rice, 99–121. Cheltenham, Gloucestershire, U.K.: Edward Elgar.

Butler, Kent, and Dowell Myers. 1984. "Boomtime in Austin, Texas: Negotiated Growth Management." *Journal of the American Planning Association* 50(4):447–458.

Byster, Leslie, and Ted Smith. 2006. "From Grassroots to Global: The Silicon Valley Toxics Coalition's Milestones in Building a Movement for Corporate Accountability and Sustainability in the High-Tech Industry." In *Challenging the Chip: Labor Rights and Environmental Justice in the Global Electronics Industry*, edited by Ted Smith, David Allan Sonnenfeld, and David N. Pellow, 111–119. Philadelphia: Temple University Press.

Campbell, Scott. 1996. "Green Cities, Growing Cities, Just Cities?" *Journal of the American Planning Association* 62(3):296–312.

Castells, Manuel. 2011. *The Rise of the Network Society*. Malden, Mass.: Wiley-Blackwell.

Castlebury, Glen. 1966. "UT to Knife Deep into East Austin." *Austin American-Statesman*, January 23.

———. 1967. "Council Shifts Boundaries of UT Renewal Project." *Austin American-Statesman*, February 3.

Chase, Lester G. 1932. *A Tabulation of City Planning Commissions in the United States*. [Rev. ed.] Washington, D.C.: Division of Building and Housing, United States National Bureau of Standards.

Chiro, Giovanna Di. 1998. "Environmental Justice from the Grassroots." In *The Struggle for Ecological Democracy*, edited by Daniel Faber, 104–136. New York: Guilford.

Christophers, Brett. 2011. "Revisiting the Urbanization of Capital." *Annals of the Association of American Geographers* 101(6):1347–1364.

Citizens Fund. 1990. *Poisons in Our Neighborhoods: Toxic Pollution in Texas*. Washington, D.C.: Citizens Fund.

Citizens' Planning Committee. 1995. "Report: Prepared for the Austin City Council." Austin: City of Austin.

———. 1996. "From Chaos to Common Ground: A Blueprint for Austin: The Citizens' Planning Committee Report to the Austin City Council." Austin: City of Austin.

———. 1999. "The Challenge for Austin's Future." Austin: City of Austin.

City of Austin. 1928. "A City Plan for Austin, Texas Prepared by Koch and Fowler." Austin: Department of Planning.

———. 1955. "The Austin Master Plan Program Prepared by Harold F. Wise and Pacific Planning and Research." Austin: City Planning Commission.

———. 1961. *The Austin Development Plan*. Austin: Department of Planning.

———. 1965. *Basic Data about Austin and Travis County*. Austin: Department of Planning.

———. 1966. Minutes of the City Council, February 10. Austin: City of Austin.

———. 1973. "Charter of the City of Austin, Article X: Planning." Austin: City of Austin.

———. 1974. *Basic Data about Austin and Travis County*. Austin: Department of Planning.

———. 1980. *Austin Tomorrow Comprehensive Plan*. Austin: Department of Planning.

———. 1986. "Charter of the City of Austin, Article X: Planning." Austin: City of Austin.

———. 1988. *Austinplan*. Austin: Planning Department.

———. 1992. "R/UDAT Implementation: A Call to Action." Austin: City of Austin. http://www.ci.austin.tx.us/downtown/rudatcall.htm.

———. 1993. "DMO Public Improvement District Resolution 930415-88 Austin City Council." Austin: City of Austin.

———. 2012. *Imagine Austin Comprehensive Plan: Vibrant, Livable, Connected Prepared by Wallace, Roberts, and Todd*. Austin: City of Austin.

Clark, Kerr. 1991. *The Great Transformation in Higher Education, 1960–1980*. Albany, N.Y.: SUNY Press.

———. 2001. *The Uses of the University*. Cambridge, Mass.: Harvard University Press.

Clark-Madison, Mike. 1998. "A City with Smarts: Austin Wising Up to Growth Plans." *Austin Chronicle*, April 17.

———. 2002. "Mapping the Changes," *Austin Chronicle*, April 5.

Clinton, William Jefferson. 1994. "Executive Order 12898—Federal Actions to Address Environmental Justice in Minority Populations and Low-Income Populations, February 11, National Archives 59 FR 7629." Washington, D.C.: United States Government Printing Office.

Cobb, James Charles. 1993. *The Selling of the South: The Southern Crusade for Industrial Development, 1936–1990*. Urbana: University of Illinois Press.

Cochrane, Allan. 2007. *Understanding Urban Policy: A Critical Approach*. Malden, Mass.: Blackwell.

Cohen, W. M., Richard Florida, and L. Randazzese. 1996. "For Knowledge and Profit: University-Industry Research Centers in the United States." Unpublished Manuscript.

Coleman, Roy. 2003. "Images from a Neoliberal City: The State, Surveillance, and Social Control." *Critical Criminology* no. 12:21–42.

Cook, W. Bruce. 1997. *From Wasteland to Wealth: The Incredible Story of the State of Texas Permanent University Fund*. Austin: University of Texas Investment Management Company.

Cooke, Phil. 2005. "Regionally Asymmetric Knowledge Capabilities and Open Innovation: Exploring 'Globalisation 2'—A New Model of Industry Organisation." *Research Policy* 34(8):1128–1149.

Cooksey, Frank. 2008. Personal Interview with former mayor, August 19.

Copelin, Laylan. 1996. "Back from the Bust." *Austin American-Statesman*, May 26.

Coppola, Sarah. 2011. "Council to Mull Districts." *Austin American-Statesman*, April 23.

Cox, Mike, and Bob Banta. 1984. "51 Rushed to Hospital after Leak at Motorola." *Austin American-Statesman*, January 25.

Cryer, Bill. 1969. "East Austinites Air Blackshear Grips." *Austin American*, May 9.

Cunningham, William H., and M. Jones. 2013. *The Texas Way: Money, Power, Politics, and Ambition at the University*. Austin: Dolph Briscoe Center for American History, University of Texas at Austin.

Curzen, Myron, and Michael Lee. 1985. "The Financial and Legal Aspects of Project Development." In *Research Parks and Other Ventures: The University/Real Estate Connection*, edited by Rachelle L. Levitt, 52–63. Washington, D.C.: The Urban Land Institute.

Cutter, Susan L. 1995. "Race, Class and Environmental Justice." *Progress in Human Geography* no. 19:111–122.

Daily Texan. 1965a. "Owners of Condemned Land Oppose Appraised Values." *Daily Texan*, October 1.

———. 1965b. "University May Get Adjacent Property." *Daily Texan*, February 12.Davidson, Chandler. 1992. "The Voting Rights Act: A Brief History." In *Controversies in Minority Voting: The Voting Rights Act in Twenty-five Year Perspective*, edited by Bernard N. Grofman and Chandler Davidson, 7–51. Washington, D.C.: Brookings Institution.

Davidson, Chandler, and George Korbel. 1981. "At-Large Elections and Minority-Group Representation: A Re-examination of Historical and Contemporary Evidence." *Journal of Politics* 43(4):982–1005.

Davidson, John. 2012. "Austin at Large." *N+1 Magazine Online*, November 23.

Davies, Christopher S. 1986. "Life at the Edge: Urban and Industrial Evolution of Texas, Frontier Wilderness—Frontier Space, 1836–1986." *Southwestern Historical Quarterly* 89(4):443–554.

Davis, Charles Edwin. 1975. "*United States v. Texas Education Agency, et al.*: The Politics of Busing." PhD diss., University of Texas at Austin.

Dicken, Peter. 2007. *Global Shift: Mapping the Changing Contours of the World Economy*. London: Sage.

Dierwechter, Yonn. 2008. *Urban Growth Management and Its Discontents: Promises, Practices, and Geopolitics in U.S. City-Regions*. New York: Palgrave Macmillan.

Dittmar, Hank. 2009. "The Human Habitat." *Ecologist*, June 1:14–18.

Dooling, Sarah. 2009. "Ecological Gentrification." *International Journal of Urban and Regional Research* 33(3):621–639.

Downing, Diane. 1983. "Thinking for the Future: The Promise of MCC." *Austin: An Official Publication of the Austin Chamber of Commerce*, August:105–110.

Dugger, Ronnie. 1974. *Our Invaded Universities: Form, Reform, and New Starts; a Nonfiction Play for Five Stages*. New York: Norton.

Dunbar, Wells. 2011. "The Single-Member Situation." *Austin Chronicle*, February 25.

Dupont, Janelle. 1969. "UT's Role in Urban Renewal to Be Source of Resentment?" *Daily Texan*, November 19.

Dworin, Diana. 1995. "A Crazy Quily of Austin Politics." *Austin American-Statesman*, May 6.

———. 1996. "Camping Ban Takes Effect Today; Downtown Merchants Laud Policy." *Austin American-Statesman*, January 15.

Dworin, Diana, and Jeff South. 1996. "Downtown Group Wants Judges to Be Tough on 'Nuisance Crimes.'" *Austin American-Statesman*, Demember 13.

Eickmann, Jo. 1959. "Expansion Means Long Walk for Teasips." *Daily Texan*, May 21.

Eisinger, Peter K. 1988. *The Rise of the Entrepreneurial State: State and Local Economic Development Policy in the United States*. Madison: University of Wisconsin Press.

———. 1995. "State Economic Development in the 1990s: Politics and Policy Learning." *Economic Development Quarterly* 9(2):146–158.

Eckel, Peter D. and Jacqueline E. King. 2007. "United States." In *International Handbook of Higher Education*, Part 1: *Global Themes and Contemporary Challenges*, edited by James J. F. Forest and Philip G. Altbach. 1035–1053. Dordrecht, The Netherlands: Springer.

Engelking, Susan. 1996. "Austin's Opportunity Economy: A Model for Collaborative Technology Development." *Annals of the New York Academy of Sciences* 798(1):29–47.

———. 1999. "Austin's Economic Growth." *Economic Development Review* 16(2):21–24.

Ernst, Robert T., Lawrence Hugg, Richard A. Crooker, and Robert L. Ayotte. 1974. "Competition and Conflict over Land Use Change in the Inner City: Institution Versus Community." *Antipode* 6(2):70–97.

Erwin, Frank. 1973. "U.T. Austin—Brackenridge Tract—Regent Erwin's Review of the History of the Tract." In *Board of Regents*. http://www.utsystem.edu/BOR/bracktract.htm.

Etzkowitz, Henry. 2003. "Research Groups as 'Quasi-Firms': The Invention of the Entrepreneurial University." *Research Policy* 32(1):109–121.

Etzkowitz, Henry, and James Dzisah. 2008. "Unity and Diversity in High-Tech Growth and Renewal: Learning from Boston and Silicon Valley." *European Planning Studies* 16(8):1009–1024.

Etzkowitz, Henry, Andrew Webster, Christiane Gebhardt, and Branca Regina Cantisano Terra. 2000. "The Future of the University and the University of the Future: Evolution of Ivory Tower to Entrepreneurial Paradigm." *Research Policy* 29(2):313–330.

Fairweather, James S. 1990. "The University's Role in Economic Development: Lessons for Academic Leaders." *Journal of the Society of Research Administrators* 22(3):5–11.

Farley, Josh, and Norman J. Glickman. 1986. "R&D as an Economic Development Strategy: The Microelectronics and Computer Technology Corporation Comes to Austin, Texas." *Journal of the American Planning Association* 52(4):407–418.

Farr, Douglas. 2008. *Sustainable Urbanism: Urban Design with Nature*. Hoboken, N.J.: Wiley.

Feagin, Joe R. 1988. *Free Enterprise City: Houston in Political-Economic Perspective*. New Brunswick, N.J.: Rutgers University Press.

Feldman, Maryann, and Shiri Breznitz. 2009. "The American Experience in University Technology Transfer." In *Learning to Compete in European Universities: From Social*

Institution to Knowledge Business, edited by Maureen McKelvey and Magnus Holmen, 161–186. Cheltenham, Gloucestershire, U.K.: Edward Elgar.

Feldman, Maryann, and Pierre Desrochers. 2003. "Research Universities and Local Economic Development: Lessons from the History of the Johns Hopkins University." *Industry and Innovation* 10(1):5–24.

Feller, Irwin. 1988. "Evaluating State Advanced Technology Programs." *Evaluation Review* 12(3):232–252.

———. 1992. "American State Governments as Models for National Science Policy." *Journal of Policy Analysis and Management* 11(2):288–309.

———. 2004. "Virtuous and Vicious Cycles in the Contributions of Public Research Universities to State Economic Development Objectives." *Economic Development Quarterly* 18(2):138–150.

Fink, Ira. 1983. "The University as a Land Developer." *Planning for Higher Education* 12(1):9–32.

Florida, Richard L. 2005. *Cities and the Creative Class*. New York: Routledge.

———. 2013. "America's Leading Metros for Venture Capital." *Atlantic Cities*, June 17.

Florida, Richard L., and Martin Kenney. 1988. "Venture Capital–Financed Innovation and Technological Change in the USA." *Research Policy* 17(3):119–137.

Fogarty, Michael, and Amit Sinha. 1999. "Why Older Regions Can't Generalize from Route 128 and Silicon Valley." In *Industrializing Knowledge: University Industry Linkages in Japan and the United States*, edited by Lewis Branscomb, Fumio Kodama, and Richard L. Florida, 473–509. Boston: MIT Press.

Foley, Neil. 1997. *The White Scourge: Mexicans, Blacks, and Poor Whites in the Cotton Culture of Central Texas*. Berkeley: University of California Press.

Fontenot, Kelli. 2012. "South Austinites Consider Single-Member Districts." *Community Impact*, October 25.

For A Better Austin. 1953. "Cutting through the Propoganda Fog!" *Austin American*, January 29.

Fosler, R. Scott. 1992. "State Economic Policy: The Emerging Paradigm." *Economic Development Quarterly* 6(1):3–13.

Frey, William H. 2012. *Population Growth in Metro America since 1980*. Washington, D.C.: The Brookings Institution.

Frontain, Michael. 2010. "Microelectronics and Computer Technology Corporation." In *Handbook of Texas Online* (http://www.tshaonline.org/handbook/online/articles /dnm01). N.p.: Texas State Historical Association.

Gandara, Ricardo. 2002. "An Evolution of East Austin." *Austin American-Statesman*, October 6.

Gearin, Elizabeth. 2004. "Smart Growth or Smart Growth Machine?" In *Up against the Sprawl: Public Policy and the Making of Southern California*, edited by Jennifer R. Wolch, Manuel Pastor, and Peter Dreier, 279–307. Minneapolis: University of Minnesota Press.

Geiger, Roger L. 1986. *To Advance Knowledge: The Growth of American Research Universities, 1900–1940*. New York: Oxford University Press.

———. 1993. *Research and Relevant Knowledge: American Research Universities since World War II*. New York: Oxford University Press.

————. 2004. *Knowledge and Money: Research Universities and the Paradox of the Marketplace.* Stanford, Calif.: Stanford University Press.

Gibbs, David, and Rob Krueger. 2007. "Containng the Contradictions of Rapid Development?" In *The Sustainable Development Paradox: Urban Political Economy in the United States and Europe,* edited by Rob Krueger and David Gibbs, 95–122. New York: Guilford.

Gibson, David V., and John Sibley Butler. 2013. "Sustaining the Technopolis: High-Technology Development in Austin, Texas, 1988–2012." *World Technopolis Association* 3(2):64–80.

Gibson, David V., and Everett Rogers. 1994. *R & D Collaboration on Trial: The Microelectronics and Computer Technology Corporation.* Boston: Harvard Business Review Press.

Gibson, David V., and Raymond W. Smilor. 1991. "The Role of the Research University in Creating and Sustaining the U.S. Technopolis." In *University Spin-off Companies: Economic Development, Faculty Entrepreneurs, and Technology Transfer,* edited by Alistair M. Brett, David V. Gibson, and Raymond W. Smilor, 31–71. Lanham, Md.: Rowman and Littlefield.

Gibson, David V., Raymond W. Smilor, and George Kozmetsky. 1991. "Austin Technology-Based Industry Report." Edited by IC2 Institute, Graduate School of Business, Greater Austin Chamber of Commerce and Ernst & Young. Austin: University of Texas at Austin.

Gibson, Timothy A. 2004. *Securing the Spectacular City: The Politics of Revitalization and Homelessness in Downtown Seattle.* Lanham, Md.: Lexington.

Glaeser, Edward. 2009. "Green Cities, Brown Suburbs." *City Journal* 19(1) (http://www.city-journal.org/2009/19_1_green-cities.html).

Glasmeier, Amy. 1988. "Factors Governing the Development of High Tech Industry Agglomerations: A Tale of Three Cities." *Regional Studies* 22(4):287–301.

Globalization and World Cities Research Network. 2012. "The World According to GaWC 2012" (http://www.lboro.ac.uk/gawc/world2012t.html).

Goldfield, David R. 1997. *Region, Race, and Cities: Interpreting the Urban South.* Baton Rouge: Louisiana State University Press.

Goldstone, Dwonna Naomi. 2006. *Integrating the 40 Acres: The Fifty-Year Struggle for Racial Equality at the University of Texas.* Athens: University of Georgia Press.

Goodspeed, Robert. 2010. "The Dilemma of Online Participation: Comprehensive Planning in Austin, Texas." Boston: MIT Department of Urban Studies and Planning.

Gotham, Kevin Fox. 2006. "The Secondary Circuit of Capital Reconsidered: Globalization and the U.S. Real Estate Sector." *American Journal of Sociology* 112(1):231–275.

Gottlieb, Paul D., and Michael Fogarty. 2003. "Educational Attainment and Metropolitan Growth." *Economic Development Quarterly* 17(4):325–336.

Gottlieb, Robert. 2005. *Forcing the Spring: The Transformation of the American Environmental Movement.* Washington, D.C.: Island Press.

Graham, Hugh Davis, and Nancy Diamond. 1997. *The Rise of American Research Universities: Elites and Challengers in the Postwar Era.* Baltimore: Johns Hopkins University Press.

Graeber, David. 2012. *The First Five Thousand Years.* Brooklyn, N.Y.: Melville House.

Greater Austin Chamber of Commerce. 2003. "Austin, Texas: Business Climate Assessment." Atlanta: Market Street Services.

Greenberg, Daniel S. 2008. *Science for Sale: The Perils, Rewards, and Delusions of Campus Capitalism*. Chicago: University of Chicago Press.

Greenberger, Scott S. 1997a. "A Legacy of Zoning Bias." *Austin American-Statesman*,
July 21.

———. 1997b. "Plan Adds Apartment Downtown; Developer Seeks City's Help to Con-
vert." *Austin American-Statesman*, February 12.

Greenhalgh, Christine, and Mark Rogers. 2010. *Innovation, Intellectual Property, and
Economic Growth*. Princeton, N.J.: Princeton University Press.

Gregor, Katherine. 2010. "Austin Comp Planning." *Austin Chronicle*, February 5.

Habermas, Jürgen. 1987. *The Philosophical Discourse of Modernity: Twelve Lectures*. Boston, Mass.: MIT Press

Hackworth, Jason. 2007. *The Neoliberal City: Governance, Ideology, and Development in
American Urbanism*. Ithaca, N.Y.: Cornell University Press.

Hall, Bronwyn H. 2002. "The Financing of Research and Development." *Oxford Review
of Economic Policy* 18(1):35–51.

Hall, Tim, and Phil Hubbard. 1998. *The Entrepreneurial City: Geographies of Politics, Regime, and Representation*. Chichester, West Sussex, England: Wiley.

Harloe, Michael, and Beth Perry. 2004. "Universities, Localities and Regional Development: The Emergence of the 'Mode 2' University?" *International Journal of Urban
and Regional Research* 28(1):212–223.

Hartenberger, Lisa, Zeynep Tufekci, and Stuart Davis. 2012. "A History of High Tech and
the Technopolis in Austin." In *Inequity in the Technopolis: Race, Class, Gender, and
the Digital Divide in Austin*, edited by J. Straubhaar, J. Spence, Z. Tufekci, and R. G.
Lentz, 63–83. Austin: University of Texas Press.

Harvey, David. 1989a. "From Managerialism to Entrepreneurialism: The Transformation
in Urban Governance in Late Capitalism." *Geografiska Annaler B* 71(1):3–17.

———. 1989b. *The Urban Experience*. Baltimore: Johns Hopkins University Press.

———. 1996. *Justice, Nature, and the Geography of Difference*. Oxford: Blackwell.

———. 1999. *The Limits to Capital*. London: Verso.

———. 2001. *Spaces of Capital: Towards a Critical Geography*. New York: Routledge.

Hazard, William, Martha Kelsey, and James Strickland. 1966. *The Climate for Renewal
in Austin, Texas: A Sociological Investigation of Attitudes toward Urban Renewal and
Community Development*. Austin: University of Texas, Community Development
Program.

Heath, William W. 1966. "Statement of W. W. Heath[,] Chairman of Regents of the
University of Texas[,] with Respect to University Expansion." Board of Regents Real
Estate Records, Dolph Briscoe Center for American History, University of Texas at
Austin.

Heimsath, Ben. 1995. "Managed Growth Important to City's Long-Term Plan." *Austin
American-Statesman*, November 13.

Held, David. 1999. *Global Transformations: Politics, Economics and Culture*. Stanford,
Calif.: Stanford University Press.

Helms, Gesa. 2008. *Towards Safe City Centres? Remaking the Spaces of an Old-Industrial
City*. Aldershot, U.K.: Ashgate.

Henderson, Jennifer A., and John J. Smith. 2002. "Academia, Industry, and the Bayh-Dole Act: An Implied Duty to Commercialize." Center for Integration of Medicine and Innovative Technology (http://www. cimit.org/news/regulatory/coi_part3.Pdf), October.

Herbert, Steve, and Elizabeth Brown. 2006. "Conceptions of Space and Crime in the Punitive Neoliberal City." *Antipode* no. 38:755–777.

Heynen, Nik. 2013. "Urban Political Ecology I: The Urban Century." *Progress in Human Geography*, August:1–7.

Hight, Bruce. 1994. "Attorney Pike Powers Files for Bankruptcy." *Austin American-Statesman*, December 2.

Hirsch, Arnold R. 1998. *Making the Second Ghetto: Race and Housing in Chicago, 1940–1960*. Chicago: University of Chicago Press.

Hobbs, Judy, and Mary Gail Rundell. 1982. "The Big Four." *Austin: An Official Publication of the Austin Chamber of Commerce*. April:47–55.

Hoch, Charles. 2007. "Making Plans: Representation and Intention." *Planning Theory* 6(1):16–35.

Holland, Richard A. 2006. "George W. Brackenrigde, George W. Littlefield, and the Shadow of the Past." In *The Texas Book: Profiles, History, and Reminiscences of the University*, edited by Richard A. Holland, 85–104. Austin: University of Texas Press.

Home Builders Association of Greater Austin. 2011. "HBA Comments on Imagine Austin Comprehensive Plan." Austin: HBA.

Home Owners' Loan Corporation. 1935. "Confidential Report of a Survey in Austin, Texas Prepared by R.L. Olson of the Mortgage Rehabilitation Division." The National Archives, Washington, D.C., Record Group 195, Austin Texas Folder.

Humphrey, David C. 2010. "Austin, Texas." In *Handbook of Texas Online* (http://www .tshaonline.org/handbook/online/articles/hda03). N.p.: Texas State Historical Association.

Jaffe, Adam. 1996. "Trends and Patterns in Research and Development Expenditures in the United States." *Proceedings of the National Academy of Sciences* 93(23):12658–12663.

Jaffe, Adam, Josh Lerner, Scott Stern, and Marie Thursby. 2007. "Academic Science and Entrepreneurship: Dual Engines of Growth?" *Journal of Economic Behavior and Organization* 63(4):573–576.

Jayson, Sharon. 1994. "City's Land Development Code to Get Renovation." *Austin American-Statesman*, December 6.

Johnson, Lyndon Baines. 1938. "Tarnish on the Violet Crown, Extension of Remarks of Hon. Morris Sheppard of Texas in the Senate of the United Slates, February 3, 1938, Appendix to the Congressional Record." Washington, D.C.: U.S. Government Printing Office.

Jonas, Andrew E. G., and David Wilson. (1999). "The Urban Growth Machine: Critical Reflections Two Decades Later." In *The Urban Growth Machine: Critical Perspectives, Two Decades Later*, edited by Andrew E. G. Jonas and David Wilson, 3–20. Albany, N.Y.: SUNY Press.

Jonas, Andrew E. G., and Aidan While. 2007. "Greening the Entrepreneurial City?" In *The Sustainable Development Paradox: Urban Political Economy in the United States and Europe*, edited by Rob Krueger and David Gibbs, 123–159. New York: Guilford.

Kahn, Terry D., and Josh Farley. 1984. "MCC Impact Assessment: Economic, Demographic and Land Use Impacts of the Microelectronic and Computer Technology Corporation." Austin: University of Texas Graduate Program in Community and Regional Planning.

Kaplan, Berry. 1983. "Houston: The Golden Buckle of the Sunbelt." In *Sunbelt Cities: Politics and Growth since World War II*, edited by Richard M. Bernard and Bradley Robert Rice, 196–212. Austin: University of Texas Press.

Kargon, Robert, Stuart W. Leslie, and Erica Schoenberger. 1992. "Far beyond Big Science: Science Regions and the Organization of Research and Development." In *Big Science: The Growth of Large-Scale Research*, edited by Peter Galison and Bruce Hevly, 334–354. Stanford: Stanford University Press.

Keatley, Anne. 1983. "Knowledge as Real Estate." *Science* no. 222:717.

Keeling, Kristin Leigh. 2000. "Austin's Struggle to Maintain and Preserve." MA thesis, University of Texas at Austin, Austin.

Keil, Roger. 2003. "Urban Political Ecology 1." *Urban Geography* 24(8):723–738.

——. 2007. "Sustaining Modernity, Modernizing Nature." In *The Sustainable Development Paradox: Urban Political Economy in the United States and Europe*, edited by Rob Krueger and David Gibbs, 41–65. New York: Guilford.

Keil, Roger, and Mark Whitehead. 2012. "Cities and the Politics of Sustainability." In *The Oxford Handbook of Urban Politics*, edited by Karen Mossberger, Susan E. Clarke, and Peter John, 520–544. Oxford, U.K.: Oxford University Press.

Kessler, Ronald. 2008. Personal Interview with former chairman of the Greater Austin Chamber of Commerce, August 26.

Kim, H. 1998. *Origins of a Technopole: The Case of Austin, Texas*. College Station: Texas A&M University Press.

Kleiner, Diana J. 2013. "Tracor." In *Handbook of Texas Online* (http://www.tshaonline. org/handbook/online/articles/dnt03). N.p.: Texas State Historical Association.

Knight, Robert. 2008. Personal Interview with downtown developer, September 25.

Kraetke, Stefan. 2012. *The Creative Capital of Cities: Interactive Knowledge Creation and the Urbanization Economies of Innovation*. Malden, Mass.: Wiley.

Kragie, Mary Elizabeth. 1986. "A Comparative Study of Selected Research Parks Implications for the Balcones Research Center." MA thesis, University of Texas at Austin, Austin.

Kraus, Steven Joseph. 1973. "Water, Sewers and Streets: The Acquisition of Public Utilities in Austin, Texas, 1875–1930." MA thesis, University of Texas at Austin, Austin.

Krueger, Rob, and David Gibbs. 2007. *The Sustainable Development Paradox: Urban Political Economy in the United States and Europe*. New York: Guilford.

Kuhlman, Martin. 1995. "Direct Action at the University of Texas during the Civil Rights Movement, 1960–1965." *Southwestern Historical Quarterly* 98(4):551–566.

Lack, Paul D. 1981. "Slavery and Vigilantism in Austin, Texas, 1840–1860." *Southwestern Historical Quarterly* 85(1):1–20.

Lacy, Sara. 1938. Typescript draft for the Federal Writer's Project Guide to Austin. In *Sectional Guides and Histories, Travis County*. Dolph Briscoe Center for American History, University of Texas at Austin.

Lambright, W. Henry, and Albert H. Tefch. 1989. "Science, Technology, and State Economic Development." *Policy Studies Journal* 18(1):135–147.

Largey, Matt. 2011. "City Council and the 'Gentleman's Agreement.'" *KUT News*, April 14.

Lawton-Smith, Helen. 2006. *Universities, Innovation and the Economy*. London: Routledge.

Lee, Yong S. 2000. "The Sustainability of University-Industry Research Collaboration: An Empirical Assessment." *Journal of Technology Transfer* 25(2):111–133.

Leicht, Kevin T., and J. Craig Jenkins. 1998. "Political Resources and Direct State Intervention: The Adoption of Public Venture Capital Programs in the American States, 1974–1990." *Social Forces* 76(4):1323–1345.

Leitner, H. 1990. "Cities in Pursuit of Economic Growth." *Political Geography* 9(2):146–170.

Lendel, Iryna. 2010. "The Impact of Research Universities on Regional Economies: The Concept of University Products." *Economic Development Quarterly* 24(3):210–230.

Leon, Marcos de. 2010. Personal Interview with former County Commissioner, April 2.

Leslie, Stuart W. 1993. *The Cold War and American Science: The Military-Industrial-Academic Complex at MIT and Stanford*. New York: Columbia University Press.

Levitt, Rachelle L., and Douglas Porter. 1985. "The University and Land Development." In *Research Parks and Other Ventures: The University/Real Estate Connection*, edited by Rachelle L. Levitt, 4–23. Washington, D.C.: Urban Land Institute.

Lewis, Rebecca, Gerrit Knaap, and Jamie Schindewolf. 2012. "The Spatial Structure of Cities in the United States." Cambridge, Mass.: Lincoln Institute of Land Policy.

Lindell, Chuck. 1995. "Dense Makes Sense to Reformers of Land Code." *Austin American-Statesman*, October 22.

———. 1996. "Director of Downtown Austin Alliance Resigns." *Austin American-Statesman*, July 19.

———. 1998. "Austin Passes All Three Bond Issues." *Austin American-Statesman*, May 3.

———. 2000. "Alliance Toasts Sprawl Fighters; sos Alliance, Others to Be Honored." *Austin American-Statesman*, January 15.

Litan, Robert E., Lesa Mitchell, and E. J. Reedy. 2008. "Commercializing University Innovations: Alternative Approaches." In *Innovation Policy and the Economy*, 8:31–57. Chicago: University of Chicago Press.

Logan, John R., and Harvey Luskin Molotch. 1987. *Urban Fortunes: The Political Economy of Place*. Berkeley: University of California Press.

Long, Joshua. 2009. "Sustaining Creativity in the Creative Archetype: The Case of Austin, Texas." *Cities* 26(4):210–219.

———. 2010. *Weird City: Sense of Place and Creative Resistance in Austin, Texas*. Austin: University of Texas Press.

———. 2014. "Constructing the Narrative of the Sustainability Fix: Sustainability, Social Justice and Representation in Austin, TX." *Urban Studies*, December 5:1–24.

Long, Walter. 1927. Personal letter dated October 25, Response to Inquiry About Austin's Planning and Government, written by John Surrett from the Kessler Plan Association. Austin History Center, Walter Long Papers Box 19, City Planning September 1926–1927.

———. 1928. Personal letter dated April 13, Regarding City Planning Commission in Austin, written by Louis Head from the *Dallas News*. Austin History Center, Walter Long Papers Box 19, City Planning 1926–1928.

————. 1948. *Something Made Austin Grow*. Austin: Austin Chamber of Commerce.

————. 1962. *From a Magnesium Plant to a Research Center*. Austin: Austin Chamber of Commerce.

————. 1964. *For All Time to Come*. Austin: Steck.

Long, Wesley, and Irwin Feller. 1972. "State Support of Research and Development: An Uncertain Path to Economic Growth." *Land Economics* 48(3):220–227.

Lopez, Soyna. 2006. "Comprehensive Planning in Austin, Texas: One Neigborhood at a Time." *Planning Forum* no. 12:53–78.

Luger, Michael. 1985. "The States and High-Technology Development." In *High Hopes for High Tech: Microelectronics Policy in North Carolina*, edited by Dale Whittington, 193–224. Chapel Hill: University of North Carolina Press.

Luger, Michael, and Harvey Goldstein. 1991. *Technology in the Garden: Research Parks and Regional Economic Development*. Chapel Hill: University of North Carolina Press.

Lyman, Ted. 1998. "Next Century Economy: Sustaining the Austin Region's Economic Advantage in the 21st Century." Austin: Greater Austin Chamber of Commerce.

Lyons, Donald, and Bill Luker Jr. 1998. "Explaining the Contemporary Spatial Structure of High-Technology Employment in Texas." *Urban Geography* 19(5): 431–458.

MacCorkle, Stuart Alexander. 1973. *Austin's Three Forms of Government*. San Antonio: Naylor.

Maddigan, Jack. 1965. "UT Area Land Dispute: Unhappy Owners Eye Regents." *Austin American*, October 1.

Mahdjoubi, Darius. 2004. "Knowledge, Innovation, and Entrepreneurship: Business Plans, Capital, Technology, and Growth of New Ventures in Austin, Texas." PhD diss., University of Texas at Austin.

Mandel, Ernest. 1975. *Late Capitalism*. London: NLB.

Marban, Alex de. 1995. "One Size Doesn't Fit All: City Land Code Revisions." *Austin Chronicle*, December 8.

Markusen, Ann. 1991. *The Rise of the Gunbelt: The Military Remapping of Industrial America*. New York: Oxford University Press.

Martin, Ken. 2012. "Poll Triggers Backlash from 10–1 Proponents." *Austin Bulldog*, October 17.

Martin, Roscoe C. 1933. "The Municipal Electorate." *Southwestern Social Science Quarterly* 14(3):193–237.

Martinez, G. A. 1993. "Legal Indeterminacy, Judicial Discretion. and the Mexican-American Litigation Experience, 1930–1980." *University of California Davis Law Review* no. 27:555–618.

Matthews, Charles Ray. 2007. "The Early Years of the Permanent University Fund from 1836 to 1937." PhD diss., University of Texas at Austin, Austin.

Mayer, Heike. 2007. "What is the Role of the University in Creating a High-Technology Region?" *Journal of Urban Technology* 14(3):33–58.

McCann, Eugene J. 2003. "Framing Space and Time in the City: Urban Policy and the Politics of Spatial and Temporal Scale." *Journal of Urban Affairs* 25(2):159–178.

————. 2007. "Inequality and Politics in the Creative City-Region: Questions of Livability and State Strategy." *International Journal of Urban and Regional Research* 31(1):188–96.

———. 2008. "Livable City/Unequal City: The Politics of Policy-Making in a 'Creative' Boomtown." *Revue Interventions économiques* 37:2–15.

McCarver, James William, Jr.. 1995. "The Blackland Miracle: An Analysis of the Development of Power in an East Austin Neighborhood from 1982 to 1994." PhD diss., University of Texas at Austin, Austin.

McDonald, Jason. 2012. *Racial Dynamics in Early Twentieth-Century Austin, Texas.* Lanham, Md.: Lexington Books.

McGown, Wayne. 1985. "The University's Role in Structuring the Development Approach." In *Research Parks and Other Ventures: The University/Real Estate Connection*, edited by Rachelle L. Levitt, 40–51. Washington, D.C.: Urban Land Institute.

McMurtry, Carol. 1963. "What's Ahead in City Plans?" *American Statesman*, October 6.

Melosi, Martin. 1983. "Dallas–Fort Worth: Marketing the Metroplex." In *Sunbelt Cities: Politics and Growth since World War II*, edited by Richard M. Bernard and Bradley Robert Rice, 162–195. Austin: University of Texas Press.

Mitchell, C. F. 1957. "Modern Planning Will Guarantee Efficient and Orderly Development." *[Austin] Statesman*, January 31.

Mitchell, Don. 1997. "The Annihilation of Space by Law: The Roots and Implications of Anti-Homeless Laws in the United States." *Antipode* 29(3):303–335.

Mitchell, Kathy. 1993. "High-Tech Wedded to Toxic Chemicals." *Austin Business Journal*, July 6.

Mixon, John. 1991. *Texas Municipal Zoning Law*. Vol. 1. Austin: Butterworth Legal Publishers.

Mollenkopf, John H. 1983. *The Contested City*. Princeton:, N.J. Princeton University Press.

Montejano, David. 2010. *Quixote's Soldiers: A Local History of the Chicano Movement, 1966–1981.* Austin: University of Texas Press.

Moody, Mary. 1973. "Proposals Could Alter City's Course." *Austin American-Statesman*, March 18.

Moore, Steven A. 2007. *Alternative Routes to the Sustainable City: Austin, Curitiba, and Frankfurt.* Lanham, Md.: Lexington Books.

Mora, Michelle. 1983. "MCC: Looking a Gift Horse in the Mouth." *Daily Texan*, June 13.

Moscoso, Eunice. 1997. "Program Lets Residents Plan Neighborhoods." *Austin American-Statesman*, April 14.

Mowery, David C., and Nathan Rosenberg. 1993. "The U.S. National Innovation System." In *National Innovation Systems: A Comparative Analysis*, edited by Richard Nelson, 29–75. New York: Oxford University Press.

Moulaert, Frank, and Farid Sekia. 2003. "Territorial Innovation Models: A Critical Survey." *Regional Studies* 37(3):289–302.

Mueller, Elizabeth J., and Sarah Dooling. 2011. "Sustainability and Vulnerability: Integrating Equity into Plans for Central City Redevelopment." *Journal of Urbanism: International Research on Placemaking and Urban Sustainability* 4(3):201–222.

Murdock, Steve H. 2013. "Population Change in Texas: Implications for Education and Economic Development." Houston: Rice University Hobby Center for the Study of Texas.

Myers, Dowell. 1984. *Quality of Life, Austin Trends 1970–1990: A Research Report of the*

Spring 1984 Course, Measuring Local Quality Of Life. Austin: Community and Regional Planning Program, University of Texas at Austin.

Nelsen, Lita L. 2005. "The Role of Research Institutions in the Formation of the Biotech Cluster in Massachusetts: The MIT Experience." *Journal of Commercial Biotechnology* 11(4):330–336.

Neuman, Michael. 2005. "The Compact City Fallacy." *Journal of Planning Education and Research* no. 25:11–26.

Newfield, Christopher. 2003. *Ivy and Industry: Business and the Making of the American University, 1880–1980.* Durham, N.C.: Duke University Press.

Noble, David. 1977. *American by Design: Science, Technology, and the Rise of Corporate Capitalism.* New York: Alfred A. Knopf.

O'Mara, Margaret Pugh. 2005. *Cities of Knowledge: Cold War Science and the Search for the Next Silicon Valley.* Princeton, N.J.: Princeton University Press.

Oc, Taner, and Steven Tiesdell. 1997. *Safer City Centres: Reviving the Public Realm.* London: Paul Chapman.

Oden, Michael. 1997. "From Assembly to Innovation: The Evolution and Current Structure of Austin's High Tech Economy." *Planning Forum* no. 3: 14–30.

Oden, Michael, Byungsu Kang, and Young Sub Kwon. 2007. "The Role of Local Amenities in the Birth and Development of High Technology Regions." Seoul, Korea: Korea Research Institute for Human Settlements.

Olson, David M. 1965. *Nonpartisan Elections: A Case Analysis.* Austin: Institute of Public Affairs, University of Texas.

Orum, Anthony M. 1987. *Power, Money, and the People: The Making of Modern Austin.* Austin: Texas Monthly.

Osborne, Jonathan, and Stephen Scheibal. 2003. "Like Go-Go 1990s, Smart Growth's Time Had Passed." *Austin American-Statesman,* June 22.

Ovetz, Robert Frank. 1996. "Entrepreneurialization, Resistance, and the Crisis of the Universities: A Case Study of the University of Texas at Austin." PhD diss., University of Texas at Austin.

Park, Andrew. 2007. "The View from Florida-Ville." *Fast Company,* March.

Parsons, Kermit C., and Georgia K. Davis. 1971. "The Urban University and Its Urban Environment." *Minerva* 9(3):361–385.

Peltz, Michael, and Marc A. Weiss. 1984. "State and Local Government Roles in Industrial Innovation." *Journal of the American Planning Association* 50(3):270–279.

Peng, Tina. 2007. "Austin, Texas." *Wall Street Journal,* August 22.

Perelman, Michael. 2003. "Intellectual Property Rights and the Commodity Form: New Dimensions in the Legislated Transfer of Surplus Value." *Review of Radical Political Economics* 35(3):304–311.

Phenix, Jann. 1996. "'New Urbanism' Can Mean Old-Fashioned." *Austin American-Statesman,* January 13.

Phillips, Michael. 2006. *White Metropolis: Race, Ethnicity, and Religion in Dallas, 1841–2001.* Austin: University of Texas Press.

Pinck, Dan. 1993. "(Re)emerging Roles for Developers: Universities as Partners." *Real Estate Finance* 10(2):62–64.

Pinkerton, James. 1984a. "Urban Removal." *Austin American-Statesman,* October 8.

———. 1984b. "Housing Lack Hold Tenants in Faded Project." *Austin American-Statesman*, October 8.

Pope, Tara. 1989. "With Austinplan, Apathy Overrides the Grand Vision." *Austin American-Statesman*, November 6.

Portney, Kent E. 2003. *Taking sustainable Cities Seriously: Economic Development, the Environment, and Quality of Life in American Cities.* Boston: MIT Press.

Powers, Pike. 2004. "Building the Austin Technology Cluster: The Role of Government and Community Collaboration in the Human Capital." In *New Governance for a New Rural Economy: Reinventing Public and Private Institutions,* edited by Nancy Novack. Mark Drabenstott, and Stephan Weiler, 53–71. Kansas City, Mo.: Federal Reserve Bank of Kansas City, Center for the Study of Rural America.

Powers, Pike. 2008. Personal Interview with former chairman of the Greater Austin Chamber of Commerce, September 11.

Premus, Robert. 1985. "The High-Tech Market for University Research Parks." In *Research Parks and Other Ventures: The University/Real Estate Connection,* edited by Rachelle L. Levitt, 24–33. Washington, D.C.: Urban Land Institute.

President's Council on Sustainable Development. 1996. "Sustainable America: A New Consensus for Prosperity, Opportunity, and a Healthy Environment for the Future." Washington, D.C.: United States Government Printing Office.

Preston, Darrell. 1998. "Austin's Powers." *The Bond Buyer,* February 17.

Prindle, David F. 1982. "Oil and the Permanent University Fund: The Early Years." *Southwestern Historical Quarterly* 86(2):277–298.

Prout, Erik. 2012. *Geography of Texas: People, Places, and Patterns.* Dubuque, Iowa: Kendall Hunt.

Ransom, Harry. 1966. Letter to Lester E. Harrell Jr. Board of Regents Real Estate Records, Dolph Briscoe Center for American History, University of Texas at Austin.

Rast, Joel. 2006. "Environmental Justice and the New Regionalism." *Journal of Planning Education and Research* 25(3):249–263.

Real Estate Council of Austin. 2011. "Real Estate Council Comments on Imagine Austin Comprehensive Plan." Austin: RECA.

Restrepo, David. 1996. "New DAA Head Promotes Downtown Housing Sector." *Austin Business Journal,* December 6, 4.

Rice, Bradley Robert. 1977. *Progressive Cities: The Commission Government Movement in America, 1901–1920.* Austin: University of Texas Press.

Rich, Meghan Ashlin. 2013. "'From Coal to Cool': The Creative Class, Social Capital, and the Revitalization of Scranton." *Journal of Urban Affairs* 35(3):365–384.

Rigby, David. 2000. "Geography and Technological Change." In *A Companion to Economic Geography,* edited by Eric Sheppard and Trevor Barnes, 202–223. Malden, Mass.: Wiley.

Rodriquez, Louis J., and Yoshikazu Fukasawa. 1996. *The Texas Economy: Twenty-First Century Economic Challenges.* Wichita Falls, Tex.: Midwestern State University Press.

Rossinow, Douglas. 1998. *The Politics of Authenticity: Liberalism, Christianity, and the New Left in America.* New York: Columbia University Press.

Rule, Charles F. 1985. "The Administration's Views on Joint Ventures." *Antitrust LJ* no. 54:1121.

Ryan, Brent D. 2011. "Reading through a Plan: A Visual Interpretation of What Plans Mean and How They Innovate." *Journal of the American Planning Association* 77(4):309–327.

Ryan, Robert H., Otis Horton, and Stanley Arbingast. 1968. *Austin, Texas R & D nucleus*. Austin: Bureau of Business Research, University of Texas at Austin.

Sadun, Lorenzo. 2012. "West Austin Democrats' November Endorsements." *Burnt Orange Report*, October 3.

Sargent, Kathy. 1989. "Austinplan Nears the Finish Line in Continuing Fight OVER Growth." *Austin Light*, March 15.

Sassen, Saskia. 2011. *Cities in a World Economy*. Thousand Oaks, Calif.: Pine Forge.

Savitch, Hank, and Paul Kantor. 2002. *Cities in the International Marketplace: The Political Economy of Urban Development in North America and Western Europe*. Princeton, N.J.: Princeton University Press.

Saxenian, AnnaLee. 1996. *Regional Advantage: Culture and Competition in Silicon Valley and Route 128*. Cambridge, Mass.: Harvard University Press.

Sayer, Andrew, and Richard Walker. 1992. *The New Social Economy: Reworking the Division of Labor*. Cambridge, Mass.: Blackwell.

Scheibal, Stephen. 2002. "The Power to Make a Difference." *Austin American-Statesman*, October 1.

Scheibal, Stephen. 2005. "Smart Growth Still Driving Development." *Austin American-Statesman*, March 20.

Schmandt, Jurgen, and Robert H. Wilson. 1988. "State Science and Technology Policies: An Assessment." *Economic Development Quarterly* 2(2):124–137.

Schmandt, Jurgen, Robert Hines Wilson, Lyndon B. Johnson School of Public Affairs, and Houston Area Research Center. 1987. *Promoting High-Technology Industry: Initiatives and Policies for State Governments*. Boulder: Westview Press.

Schoenberger, Erica. 1997. *The Cultural Crisis of the Firm*. Oxford: Blackwell.

Schwartz, Karen J. 1996a. "Austin's Downtown Architect." *Austin Business Journal*, July 6, 14.

———. 1996b. "City, DAA Begin $600,000 Renovation of Driskill Corner." *Austin Business Journal*, June 21, 4.

Scott, Allen J. 1997. "The Cultural Economy of Cities." *International Journal of Urban and Regional Research* 21(2):323–339.

Scott, Allen J., and Michael Storper. 1992. *Pathways to Industrialization and Regional Development*. New York: Routledge.

SEMATECH Foundation of Texas. 1987. "Austin's Site Proposal For Sematech's Manufacturing Research and Development Facility." In *Austin SEMATECH Proposal, Austin Chamber of Commerce*. Austin History Center.

Sevcik, Edward. 1992. "Selling the Austin Dam: A Disastrous Experiment in Encouraging Growth." *Southwestern Historical Quarterly* 96(2): 215–240.

Shahin, Jim. 1980. "City Bond Issue Draws Opposition." *Austin American-Statesman*, February 21.

Shelton, Beth Anne, Joe R. Feagin, Robert Bullard, Nestor Rodriguez, and Robert D. Thomas. 1989. *Houston: Growth and Decline in a Sunbelt Boomtown*. Philadelphia: Temple University Press.

Shermer, Elizabeth. 2011. "Sunbelt Boosterism: Industrial Recruitment, Economic Development, and Growth Politics in the Developing Sunbelt." In *Sunbelt Rising: The Politics of Place, Space, and Region*, edited by Michelle Nickerson and Darren Dochuk, 1–57. Philadelphia: University of Pennsylvania Press.

Silver, Christopher. 1997. "The Racial Origins of Zoning in American Cities." In *Urban Planning and the African American Community: In the Shadows*, edited by June Manning Thomas and Marsha Ritzdorf, 23–42. Thousand Oaks, Calif.: Sage.

Skop, Emily, and Tara Buentello. 2008. "Austin: Immigration and Transformation Deep in the Heart of Texas." In *Twenty-First Century Gateways: Immigrant Incorporation in Suburban America*, edited by Audrey Singer, Susan W. Hardwick, and Caroline B. Brettell, 257–281. Washington, D.C.: Brookings Institution Press.

Slaughter, Sheila. 1990. *The Higher Learning and High Technology: Dynamics of Higher Education Policy Formation*. Albany: SUNY Press.

Slaughter, Sheila, and Larry L. Leslie. 1997. *Academic Capitalism: Politics, Policies, and the Entrepreneurial University*. Baltimore: Johns Hopkins University Press.

Slaughter, Sheila, and Gary Rhoades. 1996. "The Emergence of a Competitiveness Research and Development Policy Coalition and the Commercialization of Academic Science and Technology." *Science, Technology & Human Values* 21(3):303–339.

———. 2004. *Academic Capitalism and the New Economy: Markets, State, and Higher Education*. Baltimore: Johns Hopkins University Press.

Slusher, Daryl. 2008. Personal Interview with former City Council member, September 5.

Smilor, Raymond, David Gibson, and Glenn Dietrich. 1990. "University Spin-out Companies: Technology Start-ups from UT-Austin." *Journal of Business Venturing* 5(1):63–76.

Smilor, Raymond, David Gibson, and George Kozmetsky. 1989. "Creating the Technopolis: High-Technology Development in Austin, Texas." *Journal of Business Venturing* 4(1):49–67.

Smilor, Raymond, Niall O'Donnell, Gregory Stein, and Robert S. Welborn. 2007. "The Research University and the Development of High-Technology Centers in the United States." *Economic Development Quarterly* 21(3):203–222.

Smith, Amy. 1992. "Sematech Stops Using Risky Liquid Chemical." *Austin Business Journal*, June 16, 1.

———. 1994. "Downtown Is Looking Up." *Austin Business Journal*, September 16, 1.

———. 2012. "Then There's This: This Isn't Portland." *Austin Chronicle*, June 22.

Smith, C. B. 1984. "Payrolls without Smokestacks by Jack Yeaman." *Texas Parade*, November 1950. In the binder titled Review of Austin Economic Development Foundation: Its Planning and Early Operations. Austin History Center.

Snow, David A., and Leon Anderson. 1993. *Down on Their Luck: A Study of Homeless Street People*. Berkeley: University of California Press.

Souther, J. Mark. 2011. "Acropolis of the Middle-West: Decay, Renewal, and Boosterism in Cleveland's University Circle." *Journal of Planning History* 10(1):30–58.

Speck, Larry. 2006. "Campus Architecture: The Heroic Decades." In *The Texas Book: Profiles, History, and Reminiscences of the University*, edited by Richard A. Holland, 125–138. Austin: University of Texas Press.

Spence, Jeremiah, Joseph D. Straubhaar, Zeynep Tufekci, Alexander Cho, and Dean Graber. 2012. "Structuring Race in the Cultural Geography of Austin." In *Inequity in the Technopolis: Race, Class, Gender, and the Digital Divide in Austin*, edited by Joseph D. Straubhaar, Jeremiah Spence, Zeynep Tufekci, and Roberta Lentz, 33–62. Austin: University of Texas Press.

Staniszewski, Frank. 1977. "Ideology and Practice in Municipal Government Reform: A Case Study of Austin." BA thesis, University of Texas at Austin.

State of Texas. 1876. Constitution of the State of Texas, Article VII, Education—The Public Free Schools.

Steiner, Frederick. 2008. "Envision Central Texas." In *Emergent Urbanism: Evolution in Urban Form, Texas*, edited by Sinclair Black, Frederick Steiner, Marisa Ballas, and Jeff Gipson, 16–19. Austin: University of Texas at Austin School of Architecture.

Stone, Clarence Nathan. 1989. *Regime Politics: Governing Atlanta, 1946–1988*. Lawrence: University Press of Kansas.

Stone, Harold A., Don Krasher Price, and Kathryn H. Stone. 1939. *City Manager Government in Austin, Texas*. Chicago: Public Administration Service.

Stone, Jackie. 2012. "City Council: Austin Is a Go." *AustinPost*, June 15.

Storper, Michael. 2009. "Roepke Lecture in Economic Geography: Regional Context and Global Trade." *Economic Geography* 85(1):1–21.

———. 2013. *Keys to the City: How Economics, Institutions, Social Interaction, and Politics Shape Development*. Princeton, N.J.: Princeton University Press.

Storper, Michael, and Allen Scott. 2009. "Rethinking Human Capital, Creativity, and Urban Growth." *Journal of Economic Geography* 9(2):147–167.

Storper, Michael, and Richard Walker. 1984. "The Spatial Division of Labor: Labor and the Location of Industries." *Sunbelt/Snowbelt: Urban Development and Regional Restructuring*, edited by Larry Sawers and William K. Tabb, 19–47. New York: Oxford University Press.

Straubhaar, Joseph D., Jeremiah Spence, Zeynep Tufekci, and Roberta Lentz. 2012. *Inequity in the Technopolis: Race, Class, Gender, and the Digital Divide in Austin*. Austin: University of Texas Press.

Swearingen, William Scott. 2010. *Environmental City: People, Place, Politics, and the Meaning of Modern Austin*. Austin: University of Texas Press.

Tang, Eric, and Chunhui Ren. 2014. *Outlier: The Case of Austin's Declining African-American Population*. Austin: University of Texas Institute for Urban Research and Policy Analysis.

Teaford, Jon C. 2000. "Urban Renewal and Its Aftermath." *Housing Policy Debate* 11(2):443–465.

Teixeira, Rodrigo Alves, and Tomas Nielsen Rotta. 2012. "Valueless Knowledge-Commodities and Financialization Productive and Financial Dimensions of Capital Autonomization." *Review of Radical Political Economics* 44(4):448–467.

Texas State Legislature. 1965. Senate Bill 187: Relating to conferring upon the Board of Regents of The University of Texas the power of eminent domain to acquire land for the use of The University of Texas system.

———. 1989. House Bill 1870: Related to a Comprehensive Plan Adopted by a Municipality. In *Section 211.004 Local Government Code*.

Thompson, Wilbur Richard. 1965. *A Preface to Urban Economics*. Washington, D.C.: Resources for the Future.

Tindall, George Brown. 1967. *The Emergence of the New South, 1913–1945*. Baton Rouge: Louisiana State University Press.

Todd, Bruce. 2008. Personal Interview with former mayor, August 29.

Toohey, Mary. 2012. "Vote on 30-Year Growth Plan Nears." *Austin American-Statesman*, June 11.

Tretter, Eliot M. 2009. "The Cultures of Capitalism." *Antipode* 41(1):111–132.

———. 2013. "Austin Restricted: Progressivism, Zoning, Private Racial Covenants, and the Making of a Segregated City." Austin: University of Texas Institute for Urban Research and Policy Analysis.

Tretter, Eliot M., and Melissa Adams. 2012. "The Privilege of Staying Dry: The Impact of Flooding and Racism on the Emergence of the 'Mexican' Ghetto in Austin's Low-East Side, 1880–1935." *Cities, Nature, and Development: The Politics and Production of Urban Vulnerabilities*, edited by Sarah Dooling and Gregory Simon, 187–205. Burlington, Vt.: Ashgate.

Troutman, Parke. 2004. "A Growth Machine's Plan B." *Journal of Urban Affairs* 26(5):611–622.

Tucci, Tony. 1984. "Council District Map Okayed on 4–3 Votes." *Austin American-Statesman*, December 15.

———. 1985. "Districting Plan Packs Drama." *Austin American-Statesman*, January 13.

Tyson, Kim. 1995. "Downtown Renewal Touted." *Austin American-Statesman*, September 9.

United States Congress. 1964. Report on Urban Renewal. Statement of William L. Slayton, Commissioner Urban Renewal Administration House and Home Finance Agency, before the Subcomitteee on Housing Committee on Banking and Currency, United States House of Representatives. November 21, 1963. Washington, D.C.: United States Goverment Printing Office.

———. 1983. Technology, Innovation and Regional Economic Development: Census of State Government Initiatives for High-Technology Industrial Development, Office of Technology Assessment. Washington, D.C.: United States Government Printing Office.

———. 1992. Federal Research: SEMATECH's Technological Progress and Proposed R&D Program: Briefing Report to the Chairman, Subcommittee on Defense, Committee on Appropriations, United States Senate, United States General Accounting Office. Washington, D.C.: Government Printing Office.

United States Court of Appeals Fifth Circuit. 1984. *Volma Overtonm v. City of Austin*. 748 F.2d 941.

United States Department of Education. 2012. *Digest of Education Statistics, 2011*. Washington, D.C.: National Center for Education Statistics.

United States Department of the Treasury. 2012. *The Economics of Higher Education: A Report Prepared by the Department of the Treasury with the Department of Education*. Washington, D.C.: Government Printing Office.

United States National Science Board. 1983. *University-Industry Research Relationships: Selected Studies*. Washington, D.C.: National Science Foundation, United States Government Printing Office.

University of Texas at Austin. 2013. The University of Texas at Austin Medical District Master Plan, Sasaki Associates, Inc.

University of Texas Board of Regents. (N.d.) Archway Property Acquisition (S.B. No. 142, 56th Legis.). Board of Regents Construction Files, Dolph Briscoe Center for American History, University of Texas at Austin.

———. 1960a. Master Plan Memo No. 2, "Main University Student Enrollments." Board of Regents Real Estate Records, Dolph Briscoe Center for American History, University of Texas at Austin.

———. 1960b. Master Plan Memo No. 3, "The Problem of Main University Space and Its Utilization." Board of Regents Real Estate Records, Dolph Briscoe Center for American History, University of Texas at Austin.

———. 1960c. Master Plan Memo No. 4, "Basic Needs for Classrooms, Teaching Laboratories, Libraries, and Offices; Analysis of The Brackenridge Move." Board of Regents Real Estate Records, Dolph Briscoe Center for American History, University of Texas at Austin.

———. 1960d. Press Release: UT Master Plan, September 24, 1960. UT Office of Public Affairs Records, Dolph Briscoe Center for American History, University of Texas at Austin.

———. 1960e. Prospect: A Platform for the University of Texas / Issued by the Board of Regents in Response to the Report of the Committee of 75. Dolph Briscoe Center for American History, University of Texas at Austin.

———. 1965a. Appraisal Report, University of Texas Projects J and D, Austin, Texas, Prepared by Jim Frederick and Harold Legge. Prepared for James Colvin, Business Manager, University of Texas. Austin History Center.

———. 1965b. Meeting Number 638, November 24, Main University: Statement on Relocation of Facilities for Intercollegiate Athletics. The Minutes of the Board of Regents of the University of Texas System.

———. 1966a. Land Acquisition by Urban Renewal. Board of Regents Real Estate Records, Dolph Briscoe Center for American History, University of Texas at Austin.

———. 1966b. Unpublished Minutes of Texas Board Regents on the Main University Further Land Acquistion. Board of Regents Real Estate Records, Dolph Briscoe Center for American History, University of Texas at Austin.

———. 1981. Meeting Number 782, December 10–11, U.T. Austin: Acquisition of Real Property. The Minutes of the Board of Regents of the University of Texas System.

———. 1983. Letter of Support to Governor Mark White from Regents Chairmen Jon Newton to Bring MCC to Texas. Board of Regents Records Construction Files, Box 6, Balcones Research Center Office, and Research Laboratory for MCC, Dolph Briscoe Center for American History, University of Texas at Austin.

———. 1984. Meeting Number 806, December 13–14, U.T. Austin Balcones Research Center Office and Research Laboratory for MCC. The Minutes of the Board of Regents of the University of Texas System.

———. 1987. Meeting Number 823, January 14, The University of Texas System Capital Improvement Program, 1985–1992, Approved in Principle Through January 1987. The Minutes of the Board of Regents of the University of Texas System.

———. 1988a. Comments, Hans Mark Chancellor's Council Executive Committee

Meeting, San Antonio, Texas, January 15–16. Board of Regents Records 2008-204/60, Folder 6, Dolph Briscoe Center for American History, University of Texas at Austin.

——. 1988b. Financing Sematech. Board of Regents Records 2008-204/60, Folder 6, Dolph Briscoe Center for American History, University of Texas at Austin.

——. 1988c. Letter to Governor Clements from Regents Chairman Jack Blanton, January 8. Board of Regents Records 2008-204/60, Folder 5, Dolph Briscoe Center for American History, University of Texas at Austin.

——. 1988d. Letter to Regents Chairman Jack Blanton from Governor William Clements, January 20. Board of Regents Records 2008-204/85, Folder 1, Dolph Briscoe Center for American History, University of Texas at Austin.

——. 1988e. Request for Legislative Appropriations: Sematech. University of Texas System. Austin: Texas State Library and Archives Commission.

——. 1988f. Sematech Construction and Bond Sale Finalized. Board of Regents Records 2008-204/85, Folder 2, Dolph Briscoe Center for American History, University of Texas at Austin.

University of Texas System Committee of 75. 1958. *The University of Texas Report of the Committee of 75, December 6, 1958*. Dolph Briscoe Center for American History, University of Texas at Austin.

Vermeulen, Michael. 1980. "The University as Landlord." *Institutional Investor* 14(5).

Vinson, Robert E. 1940. "The University Crosses the Bar." *Southwestern Historical Quarterly* 43(3):281–294.

Waldrep, Burnell. 1965. Letter to Frank Erwin relating to House Bill 500 (Enabling Legislation for University's Power of Eminent Domain). Board of Regents Real Estate Records, Dolph Briscoe Center for American History, University of Texas at Austin.

Walker, Richard A. 1985. "Is There a Service Economy? The Changing Capitalist Division of Labor." *Science and Society* 49(1):42–83.

Walker, Rob. 2000. "Momentum Began in Austin." *Money* 29(10):108.

Wallace, Aubrey. 1994. "Tackling Texas Toxics." In *Green Means: Living Gently on the Planet*, edited by Aubrey Wallace, 187–195. San Francisco: KQED.

Washburn, Jennifer. 2008. *University, Inc.: The Corporate Corruption of Higher Education*. New York: Basic.

Watson, Kirk. 2010. "Meet Kirk: Austin Mayor." http://www.kirkwatson.com/meet-kirk/austin-mayor/.

Wear, Ben. 1998. "Easy Austin Group Sits Out Bond Issue." *Austin American-Statesman*, April 27.

Webber, Michael, and David Rigby. 1996. *The Golden Age Illusion: Rethinking Postwar Capitalism*. New York: Guilford.

Weisman, Alan. 2007. *The World without Us*. New York: St. Martin's Press.

While, Aidan, Andrew Jonas, and David Gibbs. 2004. "The Environment and the Entrepreneurial City." *International Journal of Urban and Regional Research* 28(3):549–69.

Whitehead, Mark. 2007. *Spaces of Sustainability: Geographical Perspectives on the Sustainable Society*. New York: Routledge.

——. 2010. "Urban Economic Development and Environmental Sustainability." In *Cities and Economic Change: Restructuring and Dislocation in the Global Metropolis*, edited by Ronan Paddison and Tom Hutton, 196–216. London: Sage.

Whitney, Elizabeth. 1988. "How Austin Snared Sematech." *Austin American-Statesman*, July 24.

Wilson, Patricia, et al. 2006. "Understanding Ten Years of Neighborhood Planning in Austin: Findings and Lessons Learned." Austin: University of Texas at Austin.

Wilson, Robert, Lodis Rhodes, and Norman Glickman. 2007. *Community Change in East Austin*. Austin: Lyndon B. Johnson School of Public Affairs, University of Texas at Austin.

Windle, Rickie. 1994. "Waging War against Graffiti: Cleaning Up the City." *Austin Business Journal*, June 6, 1.

Wingfield, Brian, and Miriam Marcus. 2007. "America's Greenest States." *Forbes*, October 17.

Winling, LaDale. 2010. "Economic Development and the Landscape of Knowledge." *Journal of Urban History* 36(4):528–536.

———. 2011. "Students and the Second Ghetto: Federal Legislation, Urban Politics, and Campus Planning at the University of Chicago." *Journal of Planning History* 10(1):59–86.

Wright, Scott. 1994. "At-Large Elections vs. Single Member Districts." *Austin American-Statesman*, May 2.

Youtie, Jan, and Philip Shapira. 2008. "Building an Innovation Hub: A Case Study of the Transformation of University Roles in Regional Technological and Economic Development." *Research Policy* 37(8):1188–1204.

Yznaga, Mark. 2013. Personal Interview with political consultant, May 22.

———. 2010. Personal Interview with political consultant, April 6.

Zacharias, Beth. 1997. "Blue about the 'Green' City Council?" *Austin Business Journal*, July 20.

Zeller, Christian. 2007. "From the Gene to the Globe: Extracting Rents Based on Intellectual Property Monopolies." *Review of International Political Economy* 15(1):86–115.

INDEX

GEOGRAPHIES OF JUSTICE AND SOCIAL TRANSFORMATION

www.ingramcontent.com/pod-product-compliance
Lightning Source LLC
Chambersburg PA
CBHW010144270326
41928CB00018B/3245